AF679379

Modern Water Resources Engineering

Modern Water Resources Engineering

Braxton Abbot

STATES
ACADEMIC PRESS
www.statesacademicpress.com

States Academic Press,
109 South 5th Street,
Brooklyn, NY 11249, USA

Visit us on the World Wide Web at:
www.statesacademicpress.com

© States Academic Press, 2023

This book contains information obtained from authentic and highly regarded sources. All chapters are published with permission under the Creative Commons Attribution Share Alike License or equivalent. A wide variety of references are listed. Permissions and sources are indicated; for detailed attributions, please refer to the permissions page. Reasonable efforts have been made to publish reliable data and information, but the authors, editors and publisher cannot assume any responsibility for the validity of all materials or the consequences of their use.

ISBN: 978-1-63989-714-8

Trademark Notice: All trademarks used herein are the property of their respective owners. The use of any trademark in this text does not vest in the author or publisher any trademark ownership rights in such trademarks, nor does the use of such trademarks imply any affiliation with or endorsement of this book by such owners.

Cataloging-in-publication Data

Modern water resources engineering / Braxton Abbot.
p. cm.
Includes bibliographical references and index.
ISBN 978-1-63989-714-8
1. Water-supply engineering. 2. Hydraulic engineering. 3. Engineering.
4. Water resources development. I. Abbot, Braxton.

TD345 .W38 2023
627--dc23

Table of Contents

Preface

The purpose of the book is to provide a glimpse into the dynamics and to present opinions and studies of some of the scientists engaged in the development of new ideas in the field from very different standpoints. This book will prove useful to students and researchers owing to its high content quality.

Water resources engineering is a subfield of civil engineering that is involved in the study and management of the equipment, facilities and techniques that are used to manage and preserve water. The budgetary concerns, government regulations and other factors are also handled within this field while designing and constructing such equipment and systems. Water resources engineering seeks to address the complicated problems associated with groundwater, hydrology, river flow and flood plain management, sediment transport, pipe flow, hydraulic structures, water distribution, climate change, environmental studies, etc. It also plays an important role in ensuring the safety of drinking water for consumption by humans, plants and animals. This book provides significant information related to water management and engineering which will help the reader develop a good understanding of the basic concepts of modern water resources engineering. It is a complete source of knowledge on the present status of this important field.

At the end, I would like to appreciate all the efforts made by the authors in completing their chapters professionally. I express my deepest gratitude to all of them for contributing to this book by sharing their valuable works. A special thanks to my family and friends for their constant support in this journey.

Braxton Abbot

Chapter 1

Irrigation

1.1 Necessity and Importance

Irrigation is facing new challenges that require refined management and innovative design. Formerly, emphasis centered on project design. However, the current issues involve limited water supplies with several competing users, the threat of water quality degradation by excess irrigation and narrow economic margins. Meeting these challenges requires improved forecasting of irrigation water requirements.

Irrigation water requirements can be defined as the quantity or depth of irrigation water in addition to precipitation required to produce the desired crop yield and quality and to maintain an acceptable salinity balance in the root zone.

This quantity of water must be determined for uses such as irrigation scheduling for a particular field and seasonal water needs for planning, management, and development of irrigation projects. The amount and time required precipitation strongly influence irrigation water requirements.

In arid areas, annual precipitation is generally less than 10 inches and irrigation is necessary to successfully grow farm crops.

In semiarid areas, crops can be grown without irrigation but they are subject to droughts that reduce crop yields and can result in crop failure in extreme drought conditions.

Sub humid areas, which receive from 20 to 30 inches of annual precipitation are typically characterized by short dry periods.

Depending on the amount of water available, storage capacity of soils and crop rooting depth, irrigation may be required for short periods during the growing season in these areas. In humid areas, those receiving more than 30 inches of annual precipitation, it normally exceeds transpiration throughout most of the year.

However, yield decreases due to the occurrence of drought periods occur which decreases and impair quality especially for crops grown on shallow, sandy soils or that have a shallow root system. Irrigation is not needed to produce a crop in most years but may be required to protect against an occasional crop failure and to maintain product quality.

Objectives of Irrigation

Irrigation is facing new challenges that require refined management and innovative design. Formerly, emphasis centered on project design. However, the current issues involve limited water supplies with several competing users, the threat of water quality degradation by excess irrigation and narrow economic margins. Meeting these challenges requires improved forecasting of irrigation water requirements.

Irrigation water requirements can be defined as the quantity or depth of irrigation water in addition to precipitation required to produce the desired crop yield and quality and to maintain an acceptable salinity balance in the root zone.

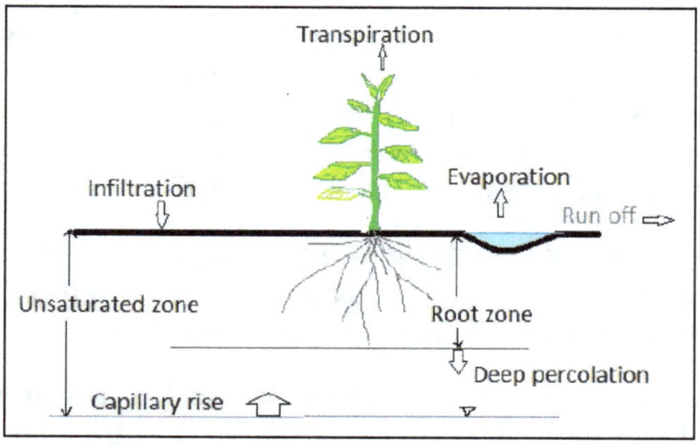

Water table.

This quantity of water must be determined for uses such as irrigation scheduling for a particular field and seasonal water needs for planning, management, and development of irrigation projects. The amount and time required precipitation strongly influence irrigation water requirements.

In arid areas, annual precipitation is generally less than 10 inches and irrigation is necessary to successfully grow farm crops.

In semiarid areas, crops can be grown without irrigation but they are subject to droughts that reduce crop yields and can result in crop failure in extreme drought conditions.

Sub humid areas, which receive from 20 to 30 inches of annual precipitation are typically characterized by short dry periods.

Depending on the amount of water available, storage capacity of soils and crop rooting depth, irrigation may be required for short periods during the growing season in these areas. In humid areas, those receiving more than 30 inches of annual precipitation, it normally exceeds transpiration throughout most of the year.

However, yield decreases due to the occurrence of drought periods occur which decreases and impair quality especially for crops grown on shallow, sandy soils or that have a shallow root system. Irrigation is not needed to produce a crop in most years but may be required to protect against an occasional crop failure and to maintain product quality.

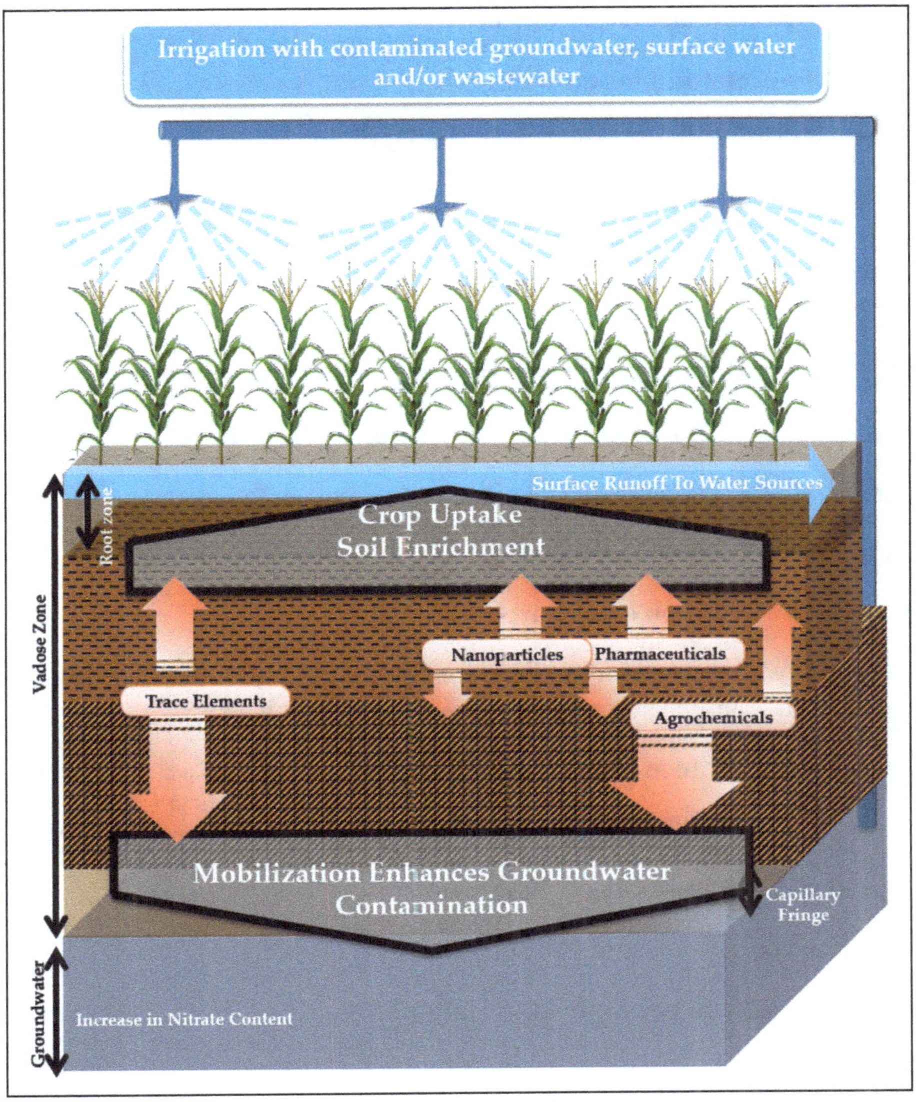

Irrigation water quality.

Importance

Water requirements for crops vary from place to place depending upon nature of crops and site. In some areas there is no need of irrigation because the conditions are fulfilled by natural resource i.e. rainfall.

Therefore the major objectives are:

- To Supply Water Partially or Totally for Crop Need.
- To Cool both the Soil and the Plant.
- To Leach Excess Salts.

- To improve Groundwater storage.
- To Facilitate continuous cropping.
- To Enhance Fertilizer Application-Fertigation.

As the population is increasing day by day and demand of more food irrigation has become necessary. If the area is in arid zone and do not have evenly distributed rainfall then this large area can be brought under cultivation because of irrigation by means of artificial application of water.

Irrigation is needed for the following purposes:

- Irrigation is required for normal growth and yield of the plants and crops.
- For metabolic processes of the plant.
- To reduce the soil temperature.
- For easy germination of seed from the soil.
- Irrigation water acts as a medium for the transport of nutrients and photosynthesis in the plant system.
- To provide insurance against short duration drought.
- To wash out and dilute the salts in the soil.
- To reduce the hazard of soil piping.
- To soften tillage pans and clods.
- Factors governing necessity of irrigation.
- Washing or diluting salts in the soil.
- Delaying bud formation by evaporative cooling.
- Promoting the function of some microorganisms.

Necessity

The following are the factors for determining the need for irrigation:

- Insufficient rainfall.
- Un-even distribution of rainfall.
- Improvement of perennial crops.
- Improving the number of crops during the year.

- Development of agriculture in desert area.
- Controlled water supply.
- Commercial crops with additional water.

Benefits of Irrigation

- During the period of low rainfall or drought yield of crops may increase or remains same due to irrigation system.
- The food production of a country will be improved by ensuring the growth of crops. This helps a country to prevent famine situation.
- Securing increased agricultural production and thus improving the nutrition of the population.
- Irrigation helps to improve the cultivation of cash crops such as vegetables, fruits, tobaccos, sugar cane.
- In some river valley projects multipurpose reservoirs are formed by constructing high dams. At these river valleys, hydroelectric power might be generated.
- Retention of water in reservoirs and possible multipurpose use thereof.
- Irrigation canal might be the source of water supply for domestic and industrial purposes.
- The reservoirs and canals will be utilized for the development of the fisher project.
- Culturing the area increasing the social and cultural level of the population.
- Recreation facilities in irrigation canals and reservoirs.
- Increases employment by providing jobs to people.
- Improvement of the micro climate. Possibility provided for waste water use and disposal.
- Improvement of water regime of the irrigated soils.

Merits of Irrigation

- During the period of low rainfall or drought yield of crops may increase or remains same due to irrigation system.
- The food production of a country will be improved by ensuring the growth of crops. This helps a country to prevent famine situation.
- Securing increased agricultural production and thus improving the nutrition of the population.

- Irrigation helps to improve the cultivation of cash crops such as vegetables, fruits, tobaccos, sugar cane.
- In some river valley projects multipurpose reservoirs are formed by constructing high dams. At these river valleys, hydroelectric power might be generated.
- Retention of water in reservoirs and possible multipurpose use thereof.
- Irrigation canal might be the source of water supply for domestic and industrial purposes.
- The reservoirs and canals will be utilized for the development of the fisher project.
- Culturing the area increasing the social and cultural level of the population.
- Recreation facilities in irrigation canals and reservoirs.
- Increases employment by providing jobs to people.
- Improvement of the micro climate. Possibility provided for waste water use and disposal.
- Improvement of water regime of the irrigated soils.

Demerits of Irrigation

- Danger of water logging and salinization of soils.
- It could change properties of water in reservoirs due to waste water use and disposal.
- Deforestation of area is to be done that is to be irrigated.
- Possible spread of diseases from certain types of surface irrigation.
- Danger of pollution of water resources by return run off from irrigation.
- New diseases caused by retention of waste water in large reservoirs.
- Due to excessive irrigation climate becomes damp and cold. Thus humidity increases which is not good for health.
- Careless irrigation might lead to retention of water and create places for breeding of mosquitoes.
- Excess of irrigation might result in raising the sub soil water table and lead to water logging of the area.

1.1.1 Principal Crops and Crop Seasons

India is top producer country of many crops. The crops can be classified based on:

- Agricultural.
- Crop seasons.
- Irrigation requirement.

Principal Crops of India

The below table gives the principal crops of India, along with crop season and average depth of water required:

Crop	Sowing time	Harvesting time	Average Delta (cm)
(A) Kharif Crop			
Rice	June-July	Oct. -Nov.	120
Makai (Maize)	June-July	Sept-Oct.	45
Bajra (Spiked Millet)	June-Aug	Sept-Oct.	30
Jowar (Great Millet)	June-July	Oct. -Nov.	30
Pulses(Mung, Moth, Urd)	June-July		30
Ground Nut	May		45
(B) Rabi Crops			
Gram	Sept-Oct.	March-April	30
Wheat	Oct. -Nov.	March-April	40
Barley	Oct. -Nov.	March-April	45
Peas	Oct. -Nov.	March-April	50
Mustard	Oct.	Feb.- March	45
Tobacco	Feb.- March	June	60
Potato	Oct.	Feb.	75
(C) Eight Months Crops			
Cotton	May-June	Dec.-April	45
(D) Perennial Crops			
Sugar-cane	Feb-March	Dec. - March	80

Agricultural

Agriculture is the main stay of our economy provides employment to about two-thirds of working population. It also contributes about a third to our national income. The contributory share in gross domestic product at present is 24.5%.

Agricultural Season

Agricultural Season.

Indian agriculture is dependent to a great extent on monsoons, soil and relief. Poor marketing and storing facilities are further problems characterizing Indian agriculture.

Agricultural Development

Agricultural production has increased several folds after independence. There are certain features which hampers balanced growth and development.

Certain changes were made in method of cropping to make our country self-sufficient such as:

- Zamidari system was abolished.
- Small land holdings were consolidated.
- Some non-traditional areas developed such as Punjab known for wheat began to grow rice.
- New varieties of high yielding seeds developed.
- Electricity provided to install their own tube-wells.
- Efforts to provide marketing and transport facilities to farmers.

Agriculture consists of the following types of crops:

- Field crops: Such as wheat, rice, maize, barley, oats, great millet, spiked millet, gram, pulses etc.
- Commercial crops: Such as sugar cane, cotton, tobacco, hemp, sugar beat etc.
- Oil seed crops: Such as mustard, ground nut, sesame, linseed, caster etc.

- Horticulture crops: Consisting of various fruit crops, various vegetable crops and flower crops
- Plantation crops: Such as tea, coffee, cocao, coconut, rubber etc.
- Forage crops: Such as fodder, grass etc.
- Miscellaneous crops: Such as medicinal crops, aromatic crops, sericulture crops, condiments and spices.

Major Crops

The crops are divided into two major groups:

- Food grains
- Non-food grains

Food Crops

Rice

Rice is India's staple food. Next to China, India is one of the most leading producers of rice in the world. Geographical conditions required are clayey soil, and standing water during growth.

Temperature to be uniformly high (above 25°C) and rainfall should be between 100 to 200 cm.

It is a kharif crop.

Rice cultivation is concentrated mainly in the Northern plains which have alluvial soils and adequate water supply. West Bengal is the leading producer of rice.

Wheat

It is one of oldest crop introduced in India from Middle East.

It is a rabi or winter crop and grown in winter season when the temperature is lesser than 20°C. Annual rainfall of 50 to 75 cm is adequate for the crop.

Loamy soils of Northern plains and black soils of Deccan are suited for wheat cultivation. Wheat is sown generally in October-November and harvested in March-April. India stands fourth in wheat production with about one-eighth of world output.

Millets

Millets are coarse grains which serve as food for a large number of people in India. They are kharif crops and grow in less rainy areas in the following order – Ragi (damp area), Jowar (moist area) and Bajra (dry area).

Millets require high temperature and less rainfall. They are the alternative to rice as rainfall decreases.

Ragi is confined to areas such as Karnataka and Tamil Nadu, Jowar in Karnataka, Andhra Pradesh, Maharashtra and Bajra in drier parts of Maharashtra, Gujarat, Rajasthan and south west Uttar Pradesh. India stands first in production of millets.

Maize

Maize, being an American crop, is a relatively new entrant and is gaining popularity because of its high yields and its easy adaptability to different soils and climatic conditions. Rich in protein and requires moderate rainfall.

Pulses

India is the largest producer and also the largest consumer of pulses .They are rich in protein .It requires moderate rainfall and verifying temperature.

Pulses include gram, arhar, moong, peas, masoor and leguminous plants. They help in restoring the fertility of the soil.

Non-food Crops

Oil Seeds: India is one of the leading producers of oil seeds in the world. Main source of edible oils. Oil seeds are used for preparing paints, varnishes, perfumes, medicines, soap etc.

Main oil seeds are groundnut (kharif crop in peninsular India), rapeseed and mustard (rabi crops in wheat belt). Additional oilseeds are seasamum (Orissa, Rajasthan, West Bengal, Tamilnadu, Maharashtra), Linseed (Madhya Pradesh, Uttar Pradesh and Maharashtra), Castor seed (Gujarat) and Cotton Seed (Maharashtra, Punjab).

India is the world's third largest producer of coconut, of which Kerala accounts for two-thirds of total production:

(Million Tonnes)			
Oilseeds	1998-99	1999-2000	(estimated) 2000-01
Groundnut	9.0	5.3	6.2
Rapessed	5.7	6.0	4.3
Soyabean	7.1	6.8	5.2
Other 6 oilseeds	2.9	2.8	2.9
Total 9 oilseeds	24.7	20.9	18.6

Sugarcane

India is known as the original land of sugarcane. It is sown just before kharif season and harvested in winter.

Sugarcane requires about 100 cm of rain. New varieties of sugar such as khandsari and gur are produced from sugarcane:

Per Hectare	Productivity in India	
Crop/Item	Per Hectare Production (in Kg.)	
	1998-99	1999-2000
Total foodgrains	1571	1697
Cereals	1778	1919
Pulses	634	630
Rice (kharif)	1619	1886
Rice (Rabi)	3073	2968
Rice (Total)	1747	1990
Wheat	2590	2755
Maize	1797	1785
Oilseeds (kharif)	992	809
Oilseeds (Rabi)	868	935
Oilseeds (Total)	944	856
Groundnut (kharif)	1174	665
Groundnut (Rabi)	1370	1389
Rape seed	870	982
Sugarcane	71	71
Cotton	224	226
Jute	1875	1995
Mesta	990	1078

Fibres

Cotton

It is a kharif crop and is known as the king among fibres. India stands fourth in the total world's production. It requires warm climate and high temperature. There are three varieties: Long staple, Medium staple and Short staple.

India mostly produces medium and short staple. Gujarat and Maharashtra are some of the leading producers of cotton.

Jute

Jute was called the golden fibre of Indian sub-continent. India ranks second in the world's production of Jute in the world after Bangladesh.

Jute is a fibre plant and it requires high temperature and rainfall above 200 cm. Grows well in flood plains or in the well-drained fertile soil.

Agricultural Production

Crop	1997-98	1998-99	1999-2000	2000-01
Rice	82.5	86.0	89.5	86.7
Wheat Coarse	66.3	71.3	75.6	70.0
Cereals	30.4	31.2	30.5	29.9
Pulses	13.0	14.9	13.4	11.7
Total food-grains	192.3	203.5	208.9	198.3
Kharif food-grains	101.6	102.8	104.9	102.0
Rabi food-grains	90.7	100.7	104.0	96.3
Oil seeds	21.3	24.7	20.9	18.6
Sugarcane	279.5	288.7	299.2	300.6
Cotton	10.9	12.3	11.6	13.2
Jute/Mesta	11.0	9.8	10.5	9.9

Plantation Crops

Tea

India is the leading producer of tea. It requires a temperature of 24°C to 30°C.

Rainfall on an average should be above 200cm. Soil should be deep fertile and well drained where water stagnation does not take place.

Production Consumption and Export of Tea

Year	Production	Consumption	Export
1996	7800	618	161.7
1997	810.6	640	203.0
1998	87 U 4	645	210.3
1999	806.0	655	190.0
2000	784.0	NA	178.0

Undulating plains of Brahmaputra and Surma Valleys of Assam is known as the house of Indian tea. Tea is one of the major foreign exchange earners of India.

Coffee

It stands second as a popular beverage in the world as well as in India. Thrives in rich and well-drained soil and it grows best in tropical highlands.

If tea belongs to north-eastern part, coffee is confined to the south-western part. Nearly 50% of Indian's production is exported to various countries.

Rubber

Rubber plantations were first established in Kerala in 1902. Needs hot-wet climatic conditions .Most of the land under rubber belongs to small land holders.

Fruits

A large variety of fruits are produced in India. This includes Bananas, Pineapples, Jack-fruits, Oranges that are grown in tropical region.

India is a leading producer in Cashew nut. Apples, Plums, Peaches, Almonds which are grown in Himachal Pradesh. Fruits are also exported. India ranks second in fruit production.

Tobacco

India is fourth largest producer and sixth larger exporter of tobacco. Crop needs freedom from frost. Tobacco needs fertile soils and heavy doses of fertilizers. Andhra pradesh and Gujarat are leading producers of tobacco.

Classification Based on Crop Seasons

Based on crop season, crops are classified as follows:

- Rabi crops or Winter Crops: These crops are sown in autumn (or October) and are harvested in spring (or March). Various crops that fall under this category are: gram, wheat, barely, peas, mustard, tobacco, linseed, potato etc.
- Kharif crops or monsoon crops: These crops are sown by the beginning of the southwest monsoon and are harvested in autumn. These consist of rice, maize, spiked millet, great millet, pulses, ground net etc.
- Perennial crops: These are the crops that require water for irrigation throughout the year. Examples of perennial crops are: sugar cane, fruits, vegetables etc.
- Eight months crops: These crops, such as cotton, require irrigation water for 8 months.

Agricultural Seasons in India

1. The Kharif Season

Crops are sown at the starting of south-west monsoon and harvested at the end of the south-west monsoon:

- Sowing Season: May to July.
- Harvesting Season: September to October.
- Important Crops: Jowar, Bajra, Rice, Maize, Cotton, Groundnut, Jute, Hemp, Tobacco etc.

2. The Rabi Season

During the growth period, the crops need cool climate and requires warm climate during the germination of seed and maturation:

- Sowing Season: October to December.
- Harvesting Season: February to April.
- Important Crops: Wheat, Barley, Mustard, Gram, Linseed, Masoor & Peas.

3. The Zaid Season

These Crops are raised throughout the year through artificial irrigation:

Zaid Rabi Crops, Zaid Kharif Crops and Classification Based on Irrigation Requirements

- Sowing Season: February to March.
- Harvesting Season: April-May.
- Important Crops: Watermelon, Toris, Cucumber & other vegetables.
- Sowing Season: August to September.
- Harvesting Season: December-January.
- Important Crops: Rice, Jowar, Rapeseed, Cotton, Oil seeds.

Based on irrigation requirements, crops can be classified as:

- Dry crops.
- Wet crops.
- Garden crops.

Dry crops are the one which do not require water for irrigation where only rain water is sufficient for their growth. Wet crops are those which cannot grow without irrigation. Garden crops require irrigation throughout the year.

TRIFED

Tribal Co-operative Marketing Development Federation of India Ltd. (TRIFED), was established in August 1987 and functioning started in 1988. Aim was to save tribal from exploitation by private traders and to offer remunerative prices for their minor forest produce.

TRIFED is an agency for collecting, processing, storing and developing all oil seed products. Agriculture ministry gives aid to TRIFED for compensating loss incurred due to price fluctuations.

NAFED

The National Agricultural Cooperative Marketing Federation of India Ltd. (NAFED), was established in cooperative secto r at national level for the marketing of agricultural products.

1.2 Methods of Application

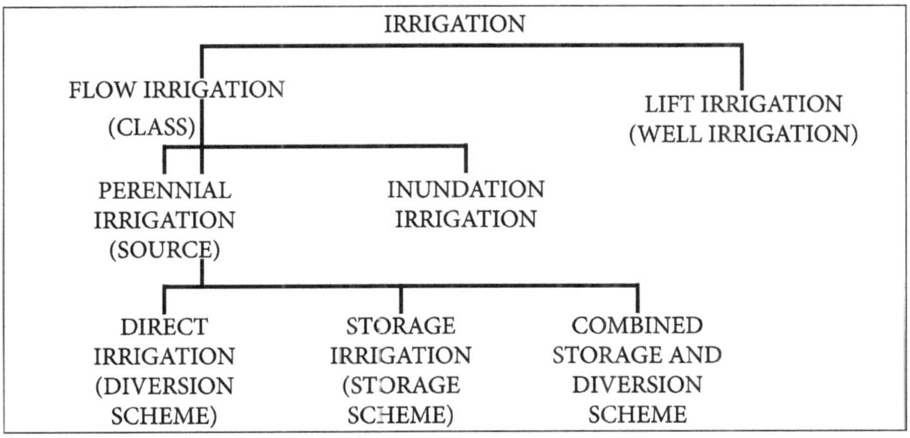

Flow Irrigation

Flow irrigation is a type of irrigation in which the supply of irrigation water available is at such a level whit is conveyed on to the land by the gravity flow. Flow irrigation may further be divided into two classes:

- Perennial irrigation system.
- Inundation or flood irrigation system.

Perennial Irrigation

In perennial irrigation system, the water required for irrigation is supplied as per the crop requirements throughout the crop period. For such a system, therefore, some storage head works, such as dams and storage weirs or barrages are required to store the excess water during floods and release it to the crops as and when it is required.

Inundation Irrigation

Inundation irrigation is carried out by deep flooding and thorough saturation of the land to be cultivated which is then drained off prior to the planting of the crop. Depending upon the source from which the water is drawn, flow irrigation can be further subdivided into three types:

- Direct irrigation (River canal irrigation): Diversion scheme.
- Storage irrigation (Reservoir or tank irrigation): Storage scheme.
- Combined storage and diversion scheme.

Direct Irrigation or River Canal Irrigation: In this type of irrigation system, water is directly diverted to the canal without attempting to store it. For such a system, diversion barrage or a low diversion weir is constructed across the river. This raises the water level in the river and thus diverts the water to the canal taking off upstream of the weir, as shown in the below figure:

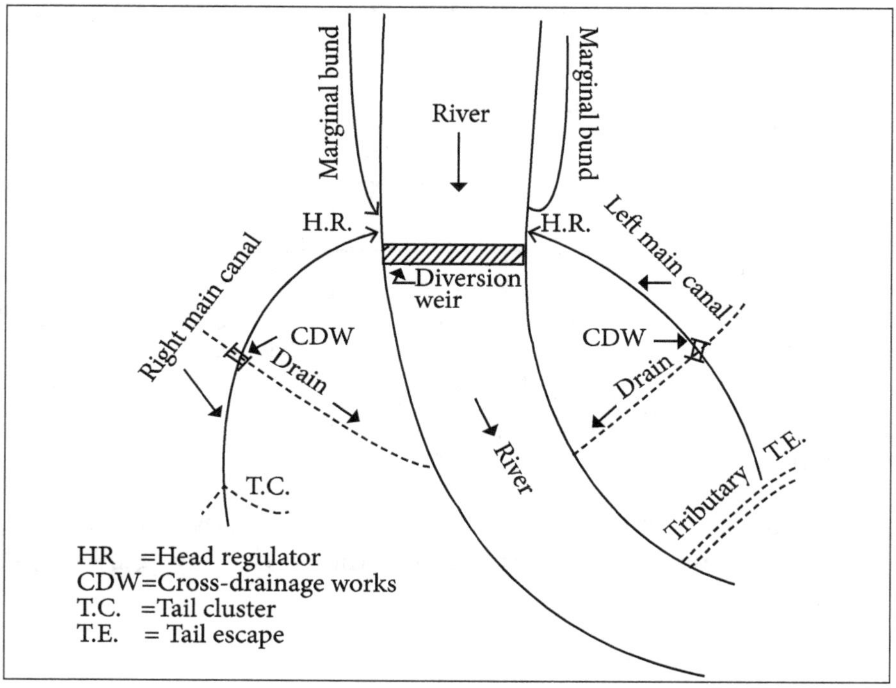

Direct Irrigation.

Usually, a direct irrigation scheme is of smaller magnitude, since there are no rigid controls over the supplies. One or two main canals may take off directly from the river. Cross-drainage works are constructed wherever distributary streams or natural drains cross the canals. In a bigger scheme, there can be a branch canal taking off from the main canal.

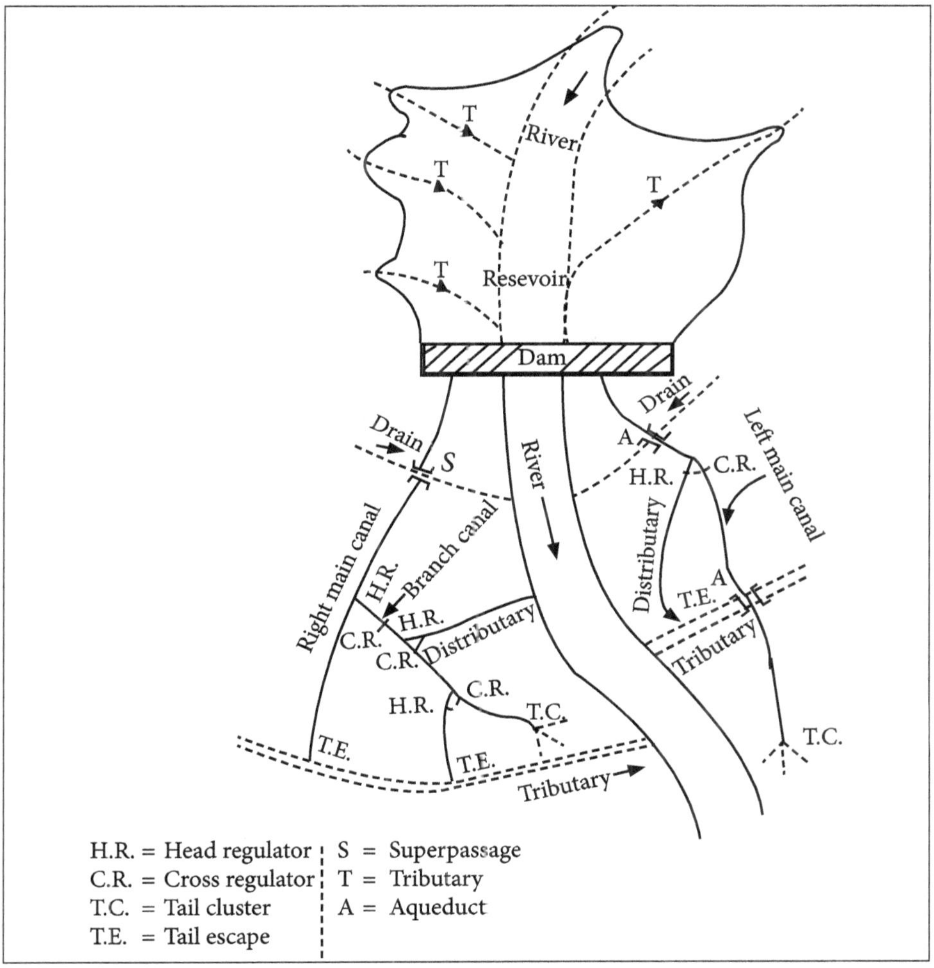

Tank Irrigation.

Storage Irrigation or Tank Irrigation: In storage irrigation system, a solid barrier, such as a dam or a storage weir is constructed across the river and water is stored in the reservoir or lake so formed. Depending upon water requirements of crops, or the hydroelectric power generation and upon the flow of water in the river, the volume of storage required is decided .The elevation storage curve for the reservoir is known from the contour plan of the basin at the site of construction. The height of the dam is then decided from the curve, corresponding to the storage-volume required. In India, most of the irrigation schemes fall under this category. Figure shows the schematic plan for such a scheme Storage irrigation scheme is comparatively of a bigger magnitude, and involve: much more expenditure than a direct irrigation scheme. One or two main canals tali off from the reservoir. Due to the formation of reservoir, certain land may be submerged to the upstream of the dam. A network of canal system convey water b the agricultural fields, through various regulatory works. Cross-drainage works suck as aqueducts, siphon aqueducts, super passages and siphons are constructed wherever natural drains cross the canals.

Combined System (Storage Cum Diversion Scheme): We have seen that in the storage irrigation system, water is stored in the reservoir, since the river is not perennial. While in case of direct irrigation system, the river is perennial and hence the water is now diverted from the river to the canal.

Sometimes, a combined scheme is adopted in which the water is first stored in the reservoir formed at the upstream side of the dam. This water is used for the water power generation. The discharge from power house is fed back into the river, to the downstream side of the dam. Thus, sufficient quantity of flow is available again in the river. At a suitable location in the downstream, a pick-up weir is constructed. This weir diverts water from the river to the canal. Figure shows such a storage-cum-diversion scheme. One typical example of such a scheme is the Kota dam and Kota barrage on Chambal river in Rajasthan.

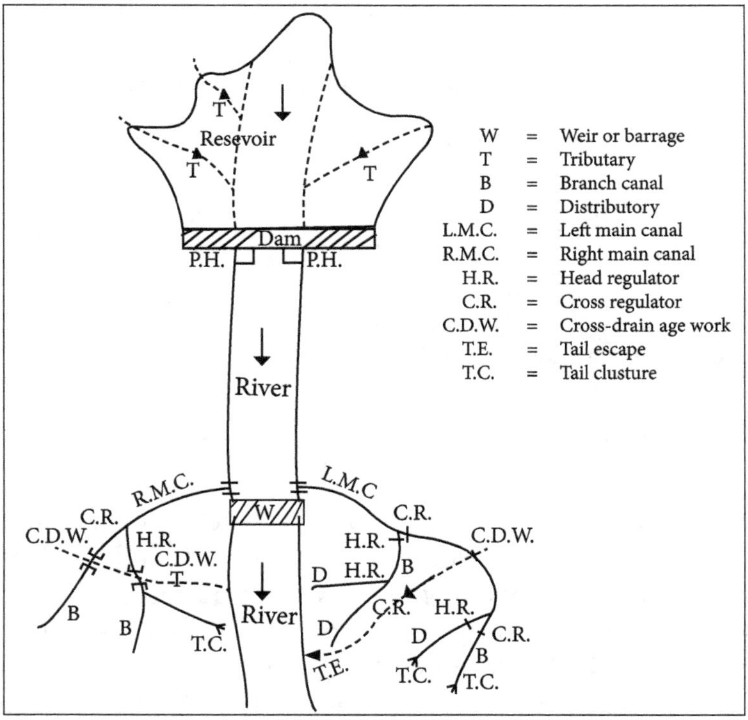

Storage cum Diversion Scheme.

Choice between the Systems Direct irrigation schemes is adopted in the circumstances where the river is perennial and has a normal flow throughout the irrigation season at any time than the requirements of the field. On the contrary, storage irrigation system is practiced when the river flow is either not perennial, or when flow is insufficient during certain parts of the crop season for irrigation requirements. In multistage river-valley development, a combined storage and diversion scheme is more useful. Lift irrigation is practiced when the water-supply is at toe low a level to run by gravitation on to the land. In such circumstances water is lifted up by mechanical means. Irrigation from wells is an example of lift irrigation, in which sub-soil water is lifted up to the surface and is then conveyed to the agricultural fields.

1.2.1 Soil, Water and plant Relationship: Soil Moisture Constants

Any given volume V of soil consists of:

- Volume of solids V_s.
- Volume of liquids V_w.
- Volume of gas V_a.

Obviously, the volume of voids $V_v = V_w + V_a$. For a fully saturated soil sample, $V_a = 0$ and $V_v = V_w$. Likewise, for a completely dry specimen, $V_w = 0$ and $V_v = V_a$. The weight of air is considered zero compared to the weights of water and soil grains. The void ratio e, the porosity n, the volumetric moisture content w, and the saturation S are defined as,

$$e = \frac{V_V}{V_S},\ n = \frac{V_V}{V},\ w = \frac{V_w}{V},\ S = \frac{V_w}{V_V}$$

Therefore,

$$w = Sn \qquad ....(1)$$

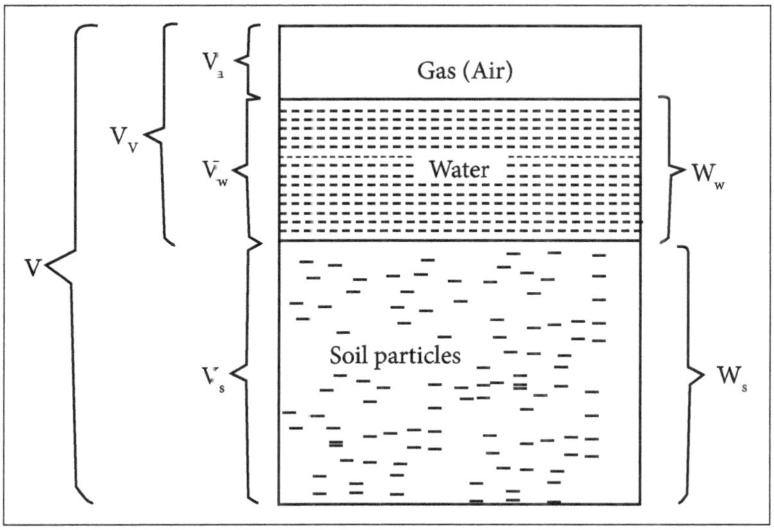

It should be noted that the value of porosity n is always less than 1.0. But, the value of void ratio e may be less, equal to, or greater than 1.0. Further, if the weight of water in a wet soil sample is W_w and the dry weight of the sample is W_s, then the dry weight moisture fraction, W is expressed as,

$$w = \frac{W_W}{W_3} \qquad ...(2)$$

The bulk density γ_b of a soil mass is the total weight of the soil per unit bulk volume, i.e.,

$$\gamma_b = \frac{W_T}{V}$$

In which,

$$W_T = W_s + W_w$$

The specific weight (or the unit weight) of the solid particles is the ratio of dry weight of the soil particles W_s to the volume of the soil particles V_s, i.e., W_s / V_s.

Thus,

$$G_b \gamma_w = \frac{W_3}{V} \qquad \text{i.e,} V = \frac{W_3}{G_b \gamma_w}$$

And,

$$G_3 \gamma_w = \frac{W_3}{V_3} \qquad \text{i.e,} V_3 = \frac{W_3}{G_3 \gamma_w}$$

Therefore,

$$\frac{V_3}{V} = \frac{G_b}{G_3} \qquad ...(3)$$

Here, γ_w is the unit weight of water and G_b and G_s are, respectively, the bulk specific gravity of soil and the relative density of soil grains.

Further,

$$1 - n = 1 - \frac{V_v}{V} = \frac{V - V_v}{V} = \frac{V_3}{V} = \frac{G_b}{G_3}$$

$$G_b = G_3 (1 - n) \qquad ...(4)$$

Also,

$$\omega \frac{V_\omega}{V} = \frac{W_\omega / \gamma_\omega}{W_3 / (G_b \gamma_\omega)} = G_b \frac{W_\omega}{W_3}$$

$$\omega = G_b\, W \quad \text{...(5)}$$

And,

$$\omega = G_3 (1 - n) W$$

Considering a soil of root-zone depth d and surface area A (i.e., bulk volume = Ad),

$$W_3 = V_3 G_3 \gamma_\omega = Ad(1-n) G_3 \gamma_\omega$$

Therefore, the dry weight moisture fraction,

$$W = \frac{W_\omega}{W_3}$$

$$= \frac{V_\omega \gamma_\omega}{Ad(1-n) G_3\, \gamma_\omega}$$

Therefore, the volume of water in the root-zone soil,

$$V_\omega = W\, Ad(1-n) G_3 \quad \text{...(6)}$$

This volume of water can also be expressed in terms of depth of water which would be obtained when this volume of water is spread over the soil surface area A.

Therefore, depth of water is, $d_\omega = \frac{V_\omega}{A}$

$$d_\omega = G_3 (1-n) Wd \quad \text{...(7)}$$

or,

$$d_\omega = \omega d \quad \text{...(8)}$$

1.3 Consumptive Use, Estimation of Consumptive Use and Crop Water Requirement

Consumptive Water Use

Consumptive water use is water removed from available provides without return to a water resource system. Evaporation from the surface of the earth into clouds of water in the air which then falls to the ground as “rain” is excluded from this model.

Crop consumptive water use is the amount of water transpired during plant growth plus what evaporates from the soil surface and foliage within the crop area.

The portion of water consumed in crop production depends on many factors especially the irrigation technology.

It is the quantity of water required by the vegetation growth of a given area. The vegetation growth includes evapotranspiration and also for the building of plant tissues plus evaporation from soils and intercepted precipitation. It is expressed in terms of depth of water. Consumptive use varying factors such as temperature, humidity, wind, speed, topography, sunlight hours, method of irrigation, moisture availability.

Mathematically,

Consumptive Use = Evapotranspiration = Evaporation + transpiration.

Factors Affecting the Consumptive Use of Water

The Consumptive use of water changes with:

- Evaporation which depends on humidity.
- Mean temperature of the month.
- Growing season of crops and the cropping pattern followed.
- Monthly precipitation over an area.
- Wind velocity in locality.
- Soil and topography.
- Irrigation practiced and method of irrigation.
- Sunlight hours.

Types of Consumptive Water Use

- Optimum Consumptive Use.
- Potential Consumptive Use.
- Seasonal Consumptive Use.

1. Optimum Consumptive Use: It is the consumptive use that produces a maximum crop yield.

2. Potential Consumptive Use: If sufficient moisture is always available to completely meet the requirements of vegetation fully covering the entire area then resulting evapotranspiration is known as Potential Consumptive Use.

3. Seasonal Consumptive Use: During entire growing season, the total amount of water consumed in the evapotranspiration by a cropped area.

1.3.1 Estimation of Consumptive Use

To measure or estimate the consumptive use the following methods are used:

- Direct Methods/Field Methods.
- Empirical Methods.
- Pan evaporation method.

1. Direct Methods

In this method field observations are made and the physical model is used for this purpose. This includes:

- Vapour Transfer Method/Soil Moisture Studies.
- Field Plot Method.
- Tanks and Lysimeter.
- Integration Method/Summation Method.
- Irrigation Method.
- Inflow Outflow Method.

Vapour Transfer Method: It is Used for estimation of water consumptive use. Soil moisture measurements are taken before and after each irrigation. The quantity of water extracted per day from soil is computed for each period.

Field Plot Method: It Select a representative plot of area and the accuracy depending upon the representativeness of plot (cropping intensity, exposure etc). It replicates the conditions of an actual sample field (field plot). Less seepage should be there.

Inflow + Rain + Outflow = Evapotranspiration

The drawback in this method is that lateral movement of water takes place although more representative to field condition. Some correction has to be applied for deep percolation as it cannot be ascertained in the field.

Tanks and Lysimeter: Here a watertight tank of cylindrical shape having diameter 2m and depth about 3m is placed vertically on the ground to measure consumptive use of water. Tank is filled with sample of soil. Bottom of the tank consists of a sand layer and a pan for collecting the surplus water.

Plants grown in the Lysimeter must be the same as in the surrounding field. Consumptive use of water is estimated by measuring the amount of water needed for satisfactory growth of plants within the tanks.

Consumptive use of water is given by,

Cu = Wa - Wd

Where,

Cu = Consuptive use of water.

Wa = Water Applied.

Wd = Water drained off.

Lysimeter studies are time consuming and expensive. Methods 1 and 2 are the more reliable methods as compare to this method.

Integration Method: In this method, it is necessary to know the division of total area, i.e. under irrigated crops, natural native vegetation area, water surface area and bare land area. Here annual consumptive use for the whole area is found in terms of volume. Expressed in Acre feet or Hectare meter.

Mathematically,

Total Evapotranspiration = Total consumptive use.

Total Area Annual Consumptive Use = Total Evapotranspiration = A + B + C + D.

Where,

A = Unit consumptive use for each crop * its area.

B = Unit consumptive use of native vegetation* its area.

C = Water surface evaporation *its area.

D = Bare land evaporation *its area.

Irrigation Method: In this method, unit consumption is multiplied by some factor. Multiplication values depend upon the type of crops in certain area. This method requires an Engineer judgment as these factors requires Engineer's investigation of certain area.

Inflow Outflow Method: For large areas annual consumptive use is found in this method.

If U is the valley consumptive use its value is given by,

U = (I + P) + (Gs – Ge) – R

Where,

U = Valley consumptive use (in acre feet or hectare meter).

I = Total inflow during a year.

P = Yearly precipitation on valley floor.

Gs = Ground Storage at the beginning of the year.

Ge = Ground Storage at the end of the year.

R = Yearly Outflow.

All the above volumes are measured in acre-feet or hectare-meter.

Empirical Methods

Empirical equations are given for the water requirement estimation. These are,

Lowry Johnson Method: The equation for this method is,

$U = 0.0015\ H + 0.9$ (Over specified).

U = Consumptive Use.

H = Accumulated degree days during the growing season calculated from maximum temperature above 32 °F.

Penman Equation: According to this method,

$U = ET = AH + 0.27\ E_a A - 0.27.$

ET = consumptive use or Evapotranspiration in mm.

E_a = Evaporation (mm/day).

H = Daily head budget at surface (mm/day).

H is a function of sunshine hours, radiation,, wind speed, vapour pressure and other climatic factors.

A = saturated vapour pressure curve slope of air at absolute temperature in °F.

Hargreave's Method: It is a very simple method.

According to this method,

$Cu = K * E_p$

Where,

Cu = Consumptive Use coefficient (varies from crop to crop).

E_p = Evapotranspiration.

K = Coefficient.

Pan Evaporation Method: It is a measurement that combines or integrates the effects of some climate elements like temperature, humidity, rain fall, drought dispersion, solar radiation, and wind.

Evaporation is to a greater extent on hot, windy, dry, sunny days and is greatly reduced when clouds block the sun and when air is cool, calm, and humid. Enable farmers and ranchers to how much water their crops may need.

1.3.2 Crop Water Requirement

It is the water required by the plants for its survival, growth, development and to produce economic parts. This requirement can be fulfilled either naturally through precipitation or artificially through irrigation.

Crop water requirement includes all losses like:

- Transpiration loss that occurs through leaves. (T).
- Evaporation loss through soil surface in cropped area (E).
- Certain application losses includes conveyance loss, percolation loss, runoff loss, etc., (WL).
- Amount of weather consumed by plants (WP) for its metabolic activities that is estimated to be less than 1% of the total water absorption. The separation of these three components are not so easy. ET loss is taken as crop water use or crop water consumptive use.
- The water needed for special purposes (WSP) like puddling operation, ploughing operation, land preparation, leaching, requirement, weeding, dissolving fertilizer and chemical, etc.

Water requirement is symbolically represented as,

$$WR = T + E + WP + WL + WSP.$$

The estimations of the water requirement of crop are one of the basic needs for crop planning on the farm and for planning of any irrigation project.

Water requirement might be defined as the quantity of water required by a crop or diversified pattern of crop in a given period of time for its normal growth under field conditions at a place.

Water requirements includes losses due to ET or CU and losses during the application of irrigation water and also the quantity of water required for special purposes such as operations like land preparation, transplanting, leaching etc.

It can be formulated as WR = ET or Cu + application loss + water for special needs. It can be stated based on "Demand" and "supply source" as follows,

$$WR = IR + ER + S$$

Where,

IR - Irrigation requirement.

ER - Effective rainfall.

S - Contribution from ground water table.

The idea about crop water requirement is essential for farm planning with respect to total quantity of water required and its efficient use for various cropping schemes of the farm or project area. Crop water requirement is also to decide the stream size and design the canal capacity.

Combined loss of evaporation and transpiration from a cropped field is called as evapo-transpiration. Otherwise known as consumptive use and denoted as ET and this is a part of water requirement,

$$CU = E + T + WP$$

Therefore,

$$WR = CU + WL + WSP$$

Crop water requirement can also be defined as water required for meeting evapotranspiration demand of the crop and special needs in case of wet land crop. It also includes other application losses both in the case of wet land and garden land crops. Also known as crop water demand.

The requirement of crop water varies from place to place, from crop to crop and depends on agro-ecological variation and crop characters.

The following features that mainly influence the crop water requirement are:

1. Crop Factors

Growth stages:

- Variety.
- Plant population.
- Duration.

- Crop growing season.

2. Soil Factors

- Structure.
- Depth.
- Texture.
- Soil chemical composition.
- Topography.

3. Climatic Factors

- Wind velocity.
- Relative humidity.
- Sunshine hours.
- Rainfall.
- Temperature.

4. Agronomic Management Factors

- Irrigation methods used.
- Frequency of irrigation and its efficiency.
- Tillage and other cultural operations like weeding, mulching etc./ inter-cropping etc.

Based on all these factors, average crop water requirement for various crops have been worked out.

1.4 Duty, Delta, Depth and Frequency

Duty and Delta

Duty: Duty represents the irrigating capacity of a unit water. It relates the area of a crop irrigated and the quantity of irrigation water required during crop growth.

For example: If 5 cumec of water is required for a crop sown in an area of 5000 hectares, The duty will be 5000/5=1000 hectares/cumec.

Delta: Delta is total depth of water required for a crop during the entire period the crop is in the field Denoted by Δ. The unit of delta is days.

Base period: It is the whole period from irrigation water first issued for preparation of the ground for planting the crop to its last watering before harvesting.

Relation between Duty and Delta:

$$\text{Delta } (\Delta) = 8.64\frac{B}{D}$$

Where,

Δ = Delta in meter.

B = Base period in days.

D = Duty in hectares/cumec.

1.4.1 Factors Affecting Duty

The duty of water of canal system depends upon a variety of the factors. The principal factors are:

- Methods and system of irrigation.
- Mode of applying water to the crops.
- Method of cultivation.
- Time and frequency of tilling.
- Type of the crop.
- Base period of the crop.
- Climatic conditions of the area.
- Quality of water.
- Method of assessment of irrigation method.
- Canal conditions.
- Character of soil and sub-soil of the canal.
- Character of soil and sub-soil of the irrigation fields.

Methods and Systems of Irrigation: In the perennial irrigation system, soil is continuously kept moist, and hence water required for initial saturation is less. Also, due to the shallow depth of the water table, deep percolation losses are less. In the inundation irrigation, there is wasteful use of water. Hence, the perennial irrigation system has more duty than the inundation irrigation. The flow irrigation system has lower duty due to the conveyance losses in the network of the canals, while the lift irrigation system has higher duty because the commanded area of each well is very near to it. Tank irrigation gives high duty due to rigid control.

Mode of Applying Water: The flood irrigation system has lesser duty than the furrow system. Sub-irrigation system gives still higher duty. The ring basin irrigation and uncontrolled flooding give less duty.

Method of Cultivation: If the land is properly ploughed and made quite loose before irrigating, the soil will have high water retention capacity in its unsaturated zone. Thus, the number of watering can, be reduced, increasing the duty. The old and conventional methods of cultivation gives less duty in comparison to the modem methods.

Time and Frequency of Tilling: Frequency of cultivation reduces the loss of moisture through weeds. Soil structure affects the plant growth to a very great extent. A good structure (i.e. the good arrangement of soil particles in relation to one another) is called good tilth of soils. When the soil is in good tilth, evaporation losses from the surface of soil is less, soil becomes properly aerated, and hence the yield of crop is also better.

Type of the Crop: The duty varies from crop to crop.

Base Period of the Crop: If the base period of the crop is more, the amount of water required will be high, hence duty will be low and vice-versa.

Climatic Conditions of the Area: The climatic conditions which affect the duty are: (i) temperature, (ii) wind, (iii) humidity, and (iv) rainfall. Due to high temperature and wind, evaporation losses will be more, and duty will be less. A humid atmosphere reduces the losses. Rainfall during the crop-period will reduce the irrigation-water requirement, and the duty will thus be higher. In this context there are two popular terms:

- Duty inclusive of rainfall, i.e. duty be considered by taking rainfall into account.
- Duty exclusive of rainfall that is the duty has been considered by not considering the rainfall in the area.

Quality of Water: If the harmful salt-content and alkali content of the water is more, water will have to be applied liberally so that the salts are leached off. This will, in turn, reduce the duty. More fertilizing matter in water will cause less consumption of water, and increase duty.

Method of Assessment of Water: Volumetric method of assessment always leads to a higher duty. This is because the farmer will use water economically. If, however, the method of assessment is based on the area under cultivation, the farmer will have a tendency to use more water, and the duty will be low.

Canal Conditions: In an earthen canal, the seepage and percolation losses will be high, resulting in the low duty. If, however, the canal is lined, the losses will be less, and duty will be high. If the high canal is so aligned that the irrigated areas are concentrated along it, the duty will be higher. The dispersion of the irrigated areas with respect to the canal tends to reduce the duty.

Character of Soil and Sub-soil of the Canal: If the canal is unlined, and if it flows through coarse grained, permeable soils, the seepage and percolation losses will be high. If the canal flows through fine grained soil, such losses will be less, and hence the duty will be higher.

Character of Soil and Sub-soil of the Irrigation Field: If the soil and sub-soil of the field is coarse-grained, percolation losses will be high. However, if there is hard pan at depth 1 to 2 metre below surface, the percolation loss reduces. The duty is also affected by the topography of the land. If the field is not level, the lower portions get more water, while higher portions may remain drier. In order to supply water to the higher places, more water will be used, and the duty will be reduced.

1.4.2 Depth and Frequency of Irrigation

The available moisture is the moisture between wilting point and the field capacity. The readily available moisture is that moisture which is easily extracted by the plants, and is approximately 75 % of the available moisture. At any time, therefore, the moisture content in the soil should be between the field capacity and the lower limit (m_o), of the readily available moisture, as shown in the below figure:

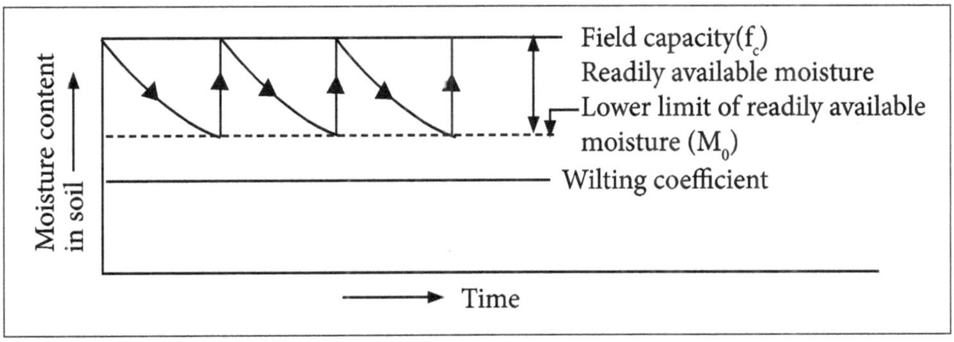

Thus, m_0 is the maximum level upto which the soil moisture may be allowed to be depleted in the root zone without fall in the yield. When watering is done, the amount of water supplied should be such that the water content is equal to the field capacity. Water will be gradually utilized consumptively by plants after the water application, and the soil moisture will start falling. When the water content in the soil reaches the value mo, fresh doses of irrigation may be done so that water content is again raised to the field capacity of the soil. The frequency of irrigation is controlled by the amount of available water contained in the root zone of the soil and the consumptive use rate. If d is the root zone depth (metres), Fe is the field capacity and mo is the lower limit of readily available moisture content, the depth of water 4 to be given during each watering is found from the following expression,

$$d_{\omega} = \frac{\gamma d}{\gamma \omega} \cdot d\left[F_C - m_o\right] \text{ metres} \qquad \text{...(1)}$$

(Where y_d is the dry unit weight of soil and y_w is the unit weight of water). Both F, and m_0 are the moisture contents to be expressed as ratio. If CL, is the daily consumptive use rate, frequency of watering f, is given by,

$$f_w = \frac{d_w}{C_u} \text{ days} \qquad \text{...(2)}$$

Time required to irrigate a certain area.

Let t be the time required to apply the desired water depth 4 to bring the water level in the soil from m_0 to the field capacity Pc, over an irrigation field of area A. If q is the discharge in the field channel, in cumecs.

We have,

$$t = \frac{A.d_w}{q} \text{ seconds} \qquad \text{...(3)}$$

Where A is the area in square metres and d,, is the depth of water to be applied in metres. If, however, area A is expressed in hectares.

We have,

$$t = \frac{d_w}{q} \cdot A \times 10^4 \text{ seconds} = \frac{2.778 \text{ Ad}_w}{q} \text{ hours} \qquad \text{...(4)}$$

1.5 Irrigation Efficiencies, Water Logging, Drainage, Standards of Quality for Irrigation Water and Crop Rotation

Irrigation efficiency is the ratio between the water stored in the soil depth inhabited with active plant roots to the water applied by the irrigation system. Thus water applied by the irrigation system and not being made available to be taken up by plant roots is wasted and reduces irrigation efficiency.

Major causes for reduced irrigation efficiency are drainage of excess irrigation water to soil layers deeper than the depth of active roots. Leakage of irrigation water to deep soil layers could result in pollution of the water table.

Irrigation efficiency of 100 percent are practically none existent even in the most modern irrigation systems.

Conservative estimates suggest that even under optimal management practices the average irrigation efficiency is estimated to be 70 percent.

Average water loss under sprinkler and drip irrigation is 30 percent but could drop to values of over 50 percent under furrow and flood irrigation.

Water losses of irrigation water under urban and landscape irrigation could easily reach 50 percent of the applied water. In Israel, a yearly saving of 300 to 400 million metric volumes of irrigation water could be saved as a result of using technologies capable of increasing substantially irrigation efficiencies.

Water Logging

In all surface water irrigation schemes, supplying the full water requirements of a crop, more water is added to the soil than is actually required to make up the deficit in the soil resulting from continuous evapotranspiration by crops. This excess water and the water that seeps into the ground from reservoirs, canals, and watercourses percolate deep into the ground to join the water table and thus, raise the water table of the area. When the rising water table reaches the root zone, the pore spaces of the root-zone soil get saturated. A land is said to be waterlogged when the pores of soil within the root zone of a plant gets saturated and the normal growth of the plant is adversely affected due to insufficient air circulation.

The depth of the water table at which it starts affecting the plant would depend on plant and soil characteristics. A land would become waterlogged sooner for deep-rooted plants than for shallow-rooted plants. Impermeable soils generally have higher capillary rise and, hence, are waterlogged more easily than permeable soils. A land is generally waterlogged when the ground water table is within 1.5 to 2.0 m below the ground surface. Water table depth is good if the water table is below 2 m and rises to 1.8 m for a period not exceeding 30 days in a year (13). If the water table is at about 1.8 m and rises to about 1.2 m for a period not exceeding 30 days in a year, the condition is considered as fair. If the water table depth is between 1.2 to 1.8 m which may rise to 0.9 m for a period not exceeding 30 days in a year, the condition of water table depth is rather poor. In a poor condition of water table depth, the water level is less than 1.2 m from the surface and is generally rising. A high water table increases the moisture content of the unsaturated surface soil and thus increases the permeability.

There may be advantages of having water table close to the surface as it may result in higher crop yield due to favourable moisture supply. This may, however, be true only for few years after water table has risen from great depths. The favourable condition may be followed by serious decrease in the crop yield in areas where alkali salts are present. With slight increase in inflow to the ground, the high water table may become too close to the ground surface and when this happens the land gets waterlogged and becomes unsuitable for cultivation.

The problem of water logging is a world-wide phenomenon which occurs mainly due to the rise of the ground water table beyond permissible limits on account of the change in ground water balance brought about by the percolation of irrigation water. It has become a problem of great importance on account of the introduction of big irrigation projects. The land subjected to waterlogging results in reduction of agricultural production. The problem of waterlogging has already affected about 5 million hectares of culturable area in India (see Table 6.2 for more details).

Causes of Waterlogging

Ground water reservoirs receive their supplies through percolation of water from the ground surface. This water may be from rainfall, from lakes or water applied to the fields for irrigation. This water percolates down to the water table and, thus, raises its position. Depending upon the elevation and the gradient of the water table, the flow may either be from surface to the ground (i.e., inflow) or ground to the surface (i.e., outflow). Outflow from a ground water reservoir includes water withdrawn through wells and water used as consumptive use. An overall balance between the inflow and outflow of a ground water reservoir will keep the water table at almost fixed level. This balance is greatly disturbed by the introduction of a canal system or a well system for irrigation. While the former tends to raise the water table, the latter tends to lower it. Waterlogging in any particular area is the result of several contributing factors. The main causes of waterlogging can be grouped into two categories:

- Natural.
- Artificial.

Natural Causes of Waterlogging

Topography, geological features, and rainfall characteristics of an area can be the natural causes of waterlogging. In steep terrain, the water is drained out quickly and, hence, chances of waterlogging are relatively low. But in flat topography, the disposal of excess water is delayed and this water stands on the ground for a longer duration. This increases the percolation of water into the ground and the chances of waterlogging. The geological features of subsoil have considerable influence on waterlogging. If the top layer of the soil is underlain by an impervious stratum, the tendency of the area getting waterlogged increases. Rainfall is the major contributing factor to the natural causes of waterlogging. Low lying basins receiving excessive rainfall have a tendency to retain water for a longer period of time and, thus get, waterlogged. Submergence of lands during floods encourages the growth of weeds and marshy grasses which obstruct the drainage of water. This, again, increases the amount of percolation of water into the ground and the chances of waterlogging.

Artificial Causes of Waterlogging

There exists a natural balance between the inflow and outflow of a ground water reservoir. This balance is greatly disturbed due to the introduction of artificial irrigation facilities. The surface reservoir water and the canal water seeping into the ground increase the inflow to the ground water reservoir. This raises the water table and the area may become waterlogged. Besides, defective method of cultivation, defective irrigation practices, and blocking of natural drainage further add to the problem of waterlogging.

Effects of Waterlogging

The crop yield is considerably reduced in a waterlogged area due to the following adverse effects of waterlogging:

- Absence of soil aeration.
- Difficulty in cultivation operations.
- Weed growth.
- Accumulation of salts. In addition, the increased dampness of the waterlogged area adversely affects the health of the persons living in that area.

Absence of Soil Aeration: In waterlogged lands, the soil pores within the root zone of crops are saturated and air circulation is cut off. Waterlogging, therefore, prevents free circulation of air in the root zone. Thus, waterlogging adversely affects the chemical processes and the bacterial activities which are essential for the proper growth of a plant. As a result, the yield of the crop is reduced considerably.

Difficulty in Cultivation: For optimum results in crop production, the land has to be prepared. The preparation of land (i.e., carrying out operations such as tillage, etc.) in wet condition is difficult and expensive. As a result, cultivation may be delayed and the crop yield adversely affected. The delayed arrival of the crop in the market brings less returns to the farmer.

Weed Growth: There are certain types of plants and grasses which grow rapidly in marshy lands. In waterlogged lands, these plants compete with the desired useful crop. Thus, the yield of the desired useful crop is adversely affected.

Accumulation of Salts: As a result of the high water table in waterlogged areas, there is an upward capillary flow of water to the land surface where water gets evaporated. The water moving upward brings with it soluble salts from salty soil layers well below the surface. These soluble salts carried by the upward moving water are left behind in the root zone when this water evaporates. The accumulation of these salts in the root zone of the soil may affect the crop yield considerably.

Remedial Measures for Waterlogging

The main cause of waterlogging in an area is the introduction of canal irrigation there. It is, therefore, better to plan the irrigation scheme in such a way that the land is prevented from getting waterlogged. Measures, such as controlling the intensity of irrigation, provision of intercepting drains, keeping the full supply level of channels as low as possible, encouraging economical use of water, removing obstructions in natural drainage, rotation of crops, running of canals by rotation, etc., can help considerably in preventing the area from getting waterlogged. In areas where the water table is relatively high, canal irrigation schemes should be planned for relatively low intensity of irrigation. In such areas canal irrigation should be allowed in the Kharif season only. Rabi irrigation should be carried out using ground water. Intercepting drains provided along canals with high embankments collect the canal water seeping through the embankments and, thus, prevent the seeping water from entering the ground. The full supply level in the channels may be kept as low as possible to reduce the seepage losses. The level should, however, be high enough to permit flow irrigation for most of the command area of the channel. For every crop there is an optimum water requirement for the maximum yield. The farmers must be made aware that the excessive use of water would harm the crop rather than benefit it. The levelling of farm land for irrigation, and a more efficient irrigation system decrease percolation to the ground and reduce the chances of waterlogging. The improvement in the existing natural drainage would reduce the amount of surface water percolating into the ground. A judicious rotation of crops can also help in reducing the chances of waterlogging. Running of canals by rotation means that the canals are run for few days and then kept dry for some days. This means that there would not be seepage for those days when the canal is dry. This, of course, is feasible only in case of distributaries and watercourses. The combined use of surface and subsurface water resources of a given area in a judicious manner to derive maximum benefits is called conjunctive use of water. During dry periods, the use of ground water is increased, and this results in lowering of the water table. The use of surface water is increased during the wet season. Because of the lowered water table, the ground water reservoir receives rainfall supplies through increased percolation. The utilisation of water resources in this manner results neither in excessive lowering of the water table nor in its excessive rising. The conjunctive use of surface and subsurface water serves as a precautionary measure against waterlogging. It helps in greater water conservation and lower evapotranspiration losses, and brings larger areas under irrigation. The most effective method of preventing waterlogging in a canal irrigated area, however, is to eliminate or reduce the seepage of canal water into the ground. This can be achieved by the lining of irrigation channels (including watercourses, if feasible). In areas which have already become waterlogged, curative methods such as surface and subsurface drainage and pumping of ground water are useful.

1.5.1 Drainage

Drainage is defined as the removal of excess water and salts from adequately irrigated agricultural lands. The deep percolation losses from properly irrigated lands and seepage from reservoirs, canals, and watercourses make drainage necessary to maintain soil productivity. Irrigation and drainage are complementary to each other. In humid areas, drainage attains much greater importance than in arid regions. Irrigated lands require adequate drainage to remain capable of producing crops. The adequate drainage of fertile lands requires the lowering of a shallow water table, and this forms the first and basic step in the reclamation of waterlogged, saline, and alkali soils. The drainage of farm lands:

- Improves soil structure and increases the soil productivity.
- Facilitates early ploughing and planting.
- Increases the depth of root zone thereby increasing the available soil moisture and plantfood.
- Increases soil ventilation.
- Increases water infiltration into the ground thereby decreasing soil erosion on the surface.
- Creates favourable conditions for growth of soil bacteria.
- Leaches excess salts from soil.
- Maintains favourable soil temperature.
- Improves sanitary and health conditions for the residents of the area.

The water table can be lowered by eliminating or controlling sources of excess water. An improvement in the natural drainage system and the provision of an artificial drainage system are of considerable help in the lowering of the water table. A natural drainage system can be properly maintained at low costs and is a feasible method of protecting irrigated lands from excessive percolation. Artificial drainage also aims at lowering the water table and is accomplished by any of the following methods:

- Open ditch drains.
- Subsurface drains.
- Drainage wells.

Open Ditch Drains

They are suitable and very often economical for surface and subsurface drainage. They permit easy entry of surface flow into the drains. Open drains are used to convey excess water to distant outlets. These accelerate the removal of storm water and thus reduce the detention time thereby decreasing the percolation of water into the ground. Open drains can be either shallow surface drains or deep open drains. Shallow surface drains do not affect subsurface drainage. Deep open drains act as outlet drains for a closed drain system and collect surface drainage too. The alignment of open drains follows the paths of natural drainage and low contours. The drains are not aligned across a pond or marshy land. Every drain has an outlet the elevation of which decides the bed and water surface elevations of the drain at maximum flow. The longitudinal slope of drain should be as large as possible and is decided on the basis of non-scouring velocities. The bed slope ranges from 0.0005 to 0.0015. Depths of about 1.5 to 3.5 m are generally adopted for open drains. The side slopes depend largely on the type of embankment soil and may vary from 1/2 H: 1V (in very stiff and compact clays) to 3H: 1V (in loose sandy formations). The open drains should be designed to carry part of storm runoff also. The cross-section of open drain is decided using the general principles of channel design. The channel will be in cutting and the height of banks will be small. If the drain has to receive both seepage and storm water, it may be desirable to have a small drain in the bed of a large open drain. This will keep the bed of the drain dry for most of the year and maintenance problems will be considerably less. Only the central deeper section will require maintenance. Open drains have the advantages of:

- Low initial cost.
- Simple construction.
- Large capacity to handle surface runoff caused by precipitation.

However, there are disadvantages too. Besides the cost of land which the open drains occupy and the need of constructing bridges across them, open drains cause:

- Difficulty in farming operations.
- Constant maintenance problems resulting from silt accumulation due to rapid weed growth in them.

Flow of clear water at low velocities permits considerable weed growth on the channel surface. The open drains have, therefore, to be cleaned frequently. In addition to manual cleaning, chemical weed killers are also used. But, at times the drain water is being used for cattles and the weed poison may be harmful to the cattles. Aquatic life is also adversely affected by the chemical weed killers.

Subsurface Drainage

It involves the creation of permanent drainage system consisting of buried pipes (or channels) which remain out of sight and, therefore, do not interfere with the farming operations. The buried drainage system can remove excess water without occupying the land area. Therefore, there is no loss of farming area. Besides, there is no weed growth and no accumulation of rubbish and, therefore, the under drainage system can remain effective for long periods with little or no need for maintenance. In some situations, however, siltation and blockage may require costly and troublesome maintenance or even complete replacement.

The materials of the buried pipes include clay pipes and concrete pipes in short lengths (permitting water entry at the joints) or long perforated and flexible plastic pipes. In addition, blankets of gravel laid in the soil, fibrous wood materials buried in the soil or such materials which can be covered by the soil and which will remain porous for long time are used for the construction of under drainage system. If such drains are to be placed in impervious soil, the drains should be surrounded by a filter of coarser material to increase the permeability and prevent migration of soil particles and blocking of drains. Mole drains are also included as subsurface drains.

The mole drains are unlined and unprotected channels of circular cross-section constructed in the subsoil at a depth of about 0.70 m by pulling a mole plough through the soil without digging a trench.

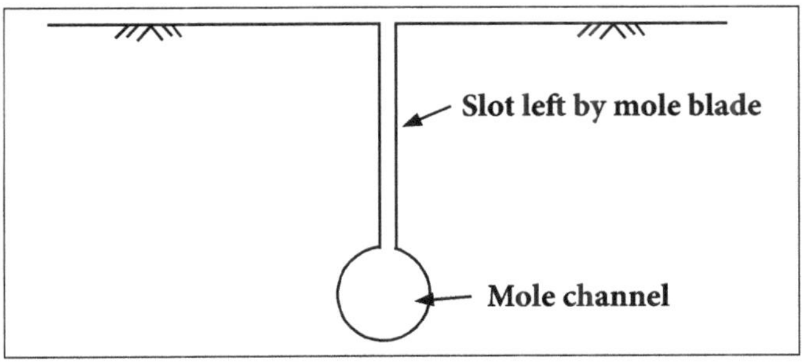

Mole drain.

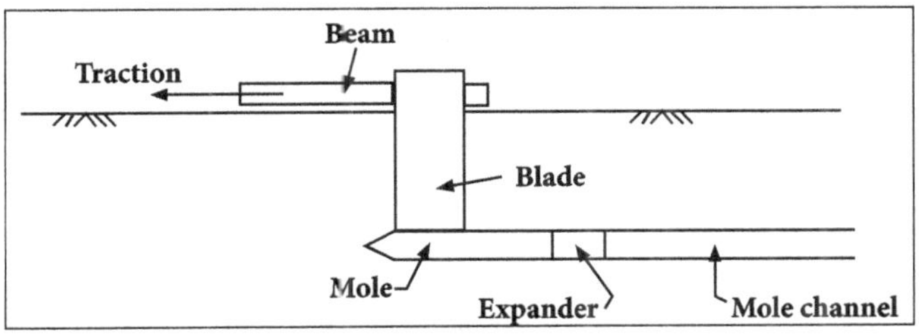

Mole plough.

The mole plough is a cylindrical metal object (about 300 to 650 mm long and 50 to 80 mm in diameter) with one of its ends bullet-shaped. The mole is attached to a horizontal beam through a thin blade as shown in Figure.

A short cylindrical metal core or sphere is attached to the rear of the mole by means of a chain. This expander helps in giving a smooth finish to the channel surface. The basic purpose of all these subsurface drains is to collect the water that flows in the subsurface region and to carry this water into an outlet channel or conveyance structure. The outlets can be either gravity outlets or pump outlets. The depth and spacing of the subsurface drains (and also deep open drains) are usually decided using Hooghoudt's equation described in the following.

Consider two drains at a spacing of B and the resulting drained water table as shown in figure below. An impermeable layer underlies the drain at a depth d. Rainfall intensity (or rate of application of irrigation water) is uniform and is equal to ra(m/s). Hooghoudt made the following assumptions to obtain a solution of the problem:

- The soil is homogeneous and isotropic.
- The hydraulic gradient at any point is equal to the slope of the water table above the point, i.e., dh/dx.
- Darcy's law is valid.

Using Darcy's law one can write,

$$q_x = ky\frac{dy}{dx} \quad ...(1)$$

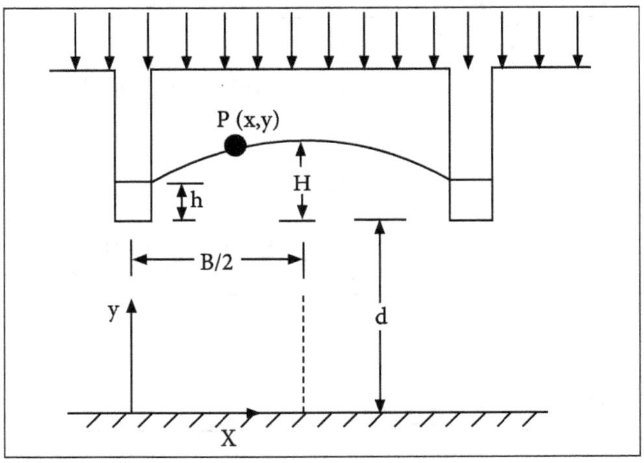

Deep open drains.

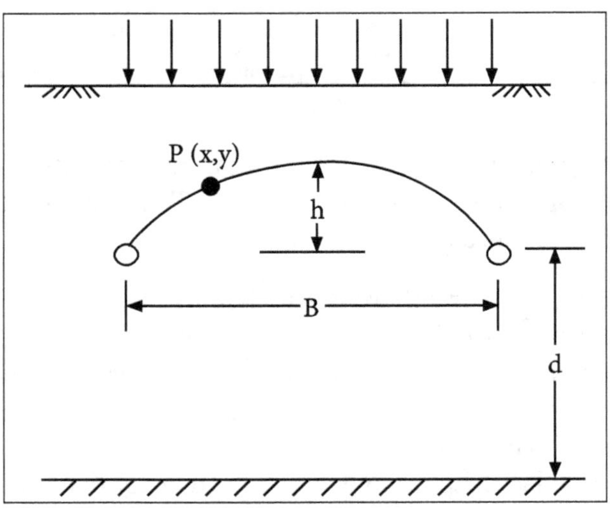

Surface open drains.

$$q_x = \left(\frac{B}{2} - x\right) r_\alpha \quad ...(2)$$

In which q_x is the discharge per unit length of drain at a section x distance away from the drain, and k is the coefficient of permeability of the soil.

Also,

Using Equation (1) and (2), one can write,

$$\left(\frac{B}{2} - x\right) r_\alpha = ky\frac{dy}{dx}$$

$$\therefore \frac{B}{2} r_\alpha dx - r_\alpha xdx = ky\,dy$$

On integrating,

$$r_\alpha \frac{B}{2} x - r_\alpha \frac{x^2}{2} = k\frac{y^2}{2} + C$$

The constant of integration C can be determined by using the boundary condition:

At x = 0, y = h + d

$$\therefore C = \frac{k(h+d)^2}{2}$$

$$r_\alpha \frac{B}{2} x - r_\alpha \frac{x^2}{2} = \frac{k}{2}\left[y^2 - (h+d)^2\right]$$

Further, at x = B/2, y = H + d

$$r_\alpha \frac{B^2}{4} - r_\alpha \frac{B^2}{8} = \frac{k}{2}\left[(H+d)^2 - (h+d)^2\right]$$

$$\text{or,} \quad r_\alpha \frac{B^2}{4} = k\left[(H+d)^2 - (h+d)^2\right]$$

$$\therefore B^2 = \frac{4k}{r} = \left[(H+d)^2 - (h+d)^2\right] \quad ...(3)$$

Equation (3) is Hooghoudt's equation for either open ditch drains or subsurface drains.

If q_d is the discharge per unit length of drain that enters the drain from two sides of the drain.

Then,

$$q_d = r_\alpha B$$

$$q_d = \frac{4k}{B}\left[(H+d)^2 - (h+d)^2\right] \qquad \text{...(4)}$$

In practice, the drain is considered empty (i.e., h = 0). Equation (4) then reduces to,

$$q_d = r_\alpha B$$

$$\therefore\ q_d = \frac{4k}{b}\left[(H+d)^2 - (h+d)^2\right] \qquad \text{...(5)}$$

$$q_d = \frac{4k\left[(H+d)^2 - (d)^2\right]}{B} \qquad \text{...(6)}$$

or,

$$q_d = \frac{4kH}{B}(H+2d)$$

$$B^2 = \frac{4kH}{r_\alpha}(H+2d) \qquad \text{...(7)}$$

Thus, knowing q_d, one can determine the spacing B of the drains. The design drain discharge (or the drainage coefficient which is defined as the amount of water that must be removed in a 24-hour period) primarily depends on the rainfall rate, size of the watershed, and the amount of surface drainage water that is admitted to the drainage system and is usually taken as equal to one per cent of average rainfall in one day.

Thus,

$$q_d = \frac{0.01 \times r_\alpha \times B}{24 \times 3600} m^3/s$$

Here, r_a is the average rainfall intensity in metres, and B is the spacing of the drains in metres. The value of B is generally between 15 and 45 m. The main drawback of the gravity drainage system is that it is not capable of lowering the water table to large depths.

Drainage wells offer a very effective method of draining an irrigated land. The soil permeability and economic considerations decide the feasibility of well drainage. Drainage wells pump water from wells drilled or already existing in the area to be drained. Design of a drainage well system will be based on established principles of well hydraulics which have been discussed in Chapter 4.

The above-mentioned remedial methods can be grouped as structural measures. In addition, the following non-structural measures can also be resorted to for preventing or reducing the menace of waterlogging:

- Adoption of tolerant crops.
- Restricting canal supplies close to crop-water needs.
- Switch over to drip irrigation.
- Conjunctive use of surface and ground water.
- Rationalization of water and power pricing policies.
- Improvement in canal irrigation management.
- Incentives for reclamation of land.

In arid regions, the bio-drainage (plantation of trees having high transpiration rates) would help in controlling the rise of ground water table and soil salinity. In addition, the bio-mass so grown acts as shelter belt in light soil area against shifting sands and dunes such as in Indira Gandhi Nahar Pariyojna (IGNP) command area in which eucalyptus trees and other trees of similar species were planted. The plantation was very effective in lowering of water table.

1.5.2 Quality Of Irrigation Water

A good irrigation water is the one which performs the above mentioned functions without any side effects which retard the plant growth. Irrigation water may be said to be unsatisfactory for its intended use if it contains (1) chemicals toxic to plants or the persons using plant as food, (2) chemicals which react with the soil to produce unsatisfactory moisture characteristics and (3) bacteria injurious to persons or animals eating plants irrigated with the water. Impurities in irrigation water The quality of irrigation water depends upon various types of impurities present in water, the following being the prominent ones: 1. Concentration of sediments in water 2. Total concentration of soluble salts (known as TDS) 3. Proportion of sodium ions to other cations 4. Concentration of toxic elements such as boron concentration 5. Concentration of bicarbonate in relation to the concentration of calcium and magnesium. 6. Bacterial concentration. The effect of concentration of sediments transported by irrigation water depends upon the type of soil of irrigation fields. If the field contains sandy soil, its fertility is improved by the deposition of sediments carried by irrigation water. However, if the sediment has been derived from eroded areas, if may reduce the fertility and decrease the permeability of soil. High concentration of salts in water may be harmful in the long run. Similarly, if water contains high concentration of sodium ions, the irrigation soils become plastic and sticky when wet and become prone to form clods and they crust on drying. Similarly high percentage of toxic elements may be harmful to certain plants such as nuts, citrus fruits and deciduous plants. High concentration of bicarbonate ions may result in precipitation of calcium and magnesium bicarbonates from the soil solution, increasing the relative proportion of sodium irons and causing sodium hazards. Bacterial contamination may not be directly harmful to the plants as such, but it may be injurious to persons or animals eating these plants.

1.5.3 Crop Rotation

The term rotation of crops is somewhat self explanatory. It implies that nature of the crop sown in a particular field is changed year after year. Necessity for rotation The necessity for rotation arises from the fact that soil gradually losses its fertility if the same crop is sown every year and the field has to be allowed to lie fallow in order to regain its fertility. The effect of fallow is obtained by rotation of crops. All crops require similar type of nutrients for their growth but all of them do not take in the same quantities or proportions; some crop favours certain plant nutrients and take them more than the others. Thus if a particular crop is grown year after year. the soil gets deficient in the plant food favoured by that crop. If different crops are to be raised there would certainly be more balanced fooding and soil deficient in one particular type of nutrient is allowed to recoupe. Crop diseases and insect pests will multiply at an alarming rate, if the same crop is to be grown continuously. Rotation will check this disease.

A leguminous crop (such as gram) if introduced in rotation will increase nitrogen content of the soil thus increasing its fertility. There could be deep rooted crops and shallow rooted crops in rotation and so they shall draw their food from different depths of soil. The soil will thus be better utilised. Rotation of crops in the long run is beneficial to the farmer and soil both as there would be a rotation of cash crops. fooder crops and soil renovating crops. The following rotations have been found useful:

- Introduction of groundnut in the usual cotton jowar rotation as per the experiments in Central Province, Hyderabad.
- In Northern India, cotton is known to have improved when it follows berseen on fields manured with phosphate.
- In Madras, Bajra when rotated with cotton has proved to be superior to Jowar being rotated with cotton.
- Cotton gives good result after gram or fallow. (v) Wheat does well after-sugar-cane and gowara. In general the following rotation of the crops give good results:

- Wheat-great millet-gram.
- Rice-gram.
- Cotton-wheat-gram.
- Cotton-wheat-sugarcane.

The reason of being one in various rotation combinations is its being leguminous crop giving nitrogen to the field.

Chapter 2

Canals

2.1 Classification and Design of Non-erodible Canals

Irrigation Canals

The canals used for irrigation purpose are known as irrigation canals. Canals may be defined as artificial channels constructed on the ground to carry the water from one place to the other. The canals may be classified as alluvial or non-alluvial canal based on the nature of source of supply. It may be termed as inundation or a permanent canal depending on how the water is fed from one system to another.

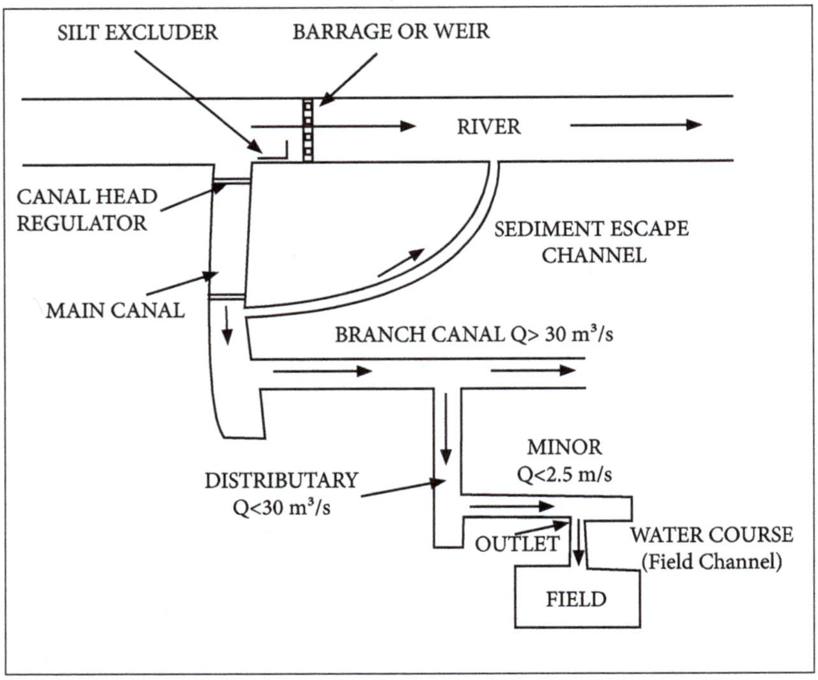

Layout of canal.

An irrigation canal system consists of canals of different sizes and capacities. Accordingly, the canals are also classified as:

- Main canal.
- Branch canal.

- Major distributary.
- Minor distributary.
- Watercourse.

Main Canal: Main Canal takes off directly from the upstream side of weir head works or dam. Usually no direct cultivation is proposed.

Main canal.

Branch Canal: All offtakes from main canal with head discharge of 14-15 cumecs and above are termed as branch canals. Acts as feeder channel for major distributaries.

Branch canal.

Major Distributary: All offtakes from main canal or branch canal with head discharge from 0.25 to 15 cumecs are called as the major distributaries.

Minor Distributary: All offtakes taking off from a major distributary getting discharge less than 0.25 cumec are termed as minor distributaries

Water Course: Small channels which carry water from outlets of a major or minor distributary or a branch canal to the fields to be irrigated.

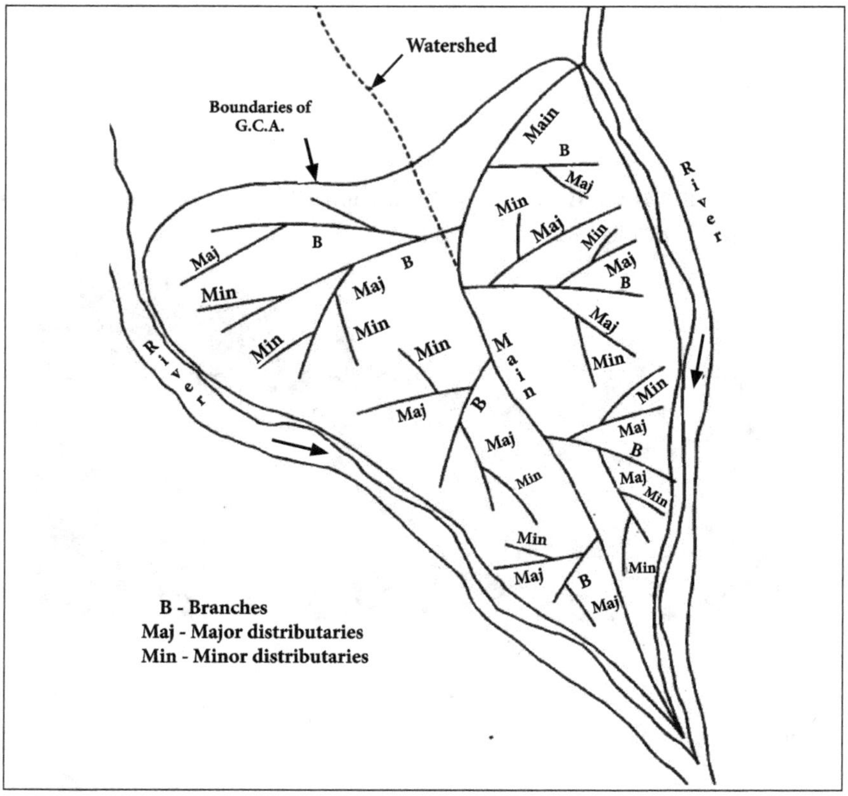

Distributary

Classification of Canal

Alluvial Canal: A canal flowing through alluvium soil is called an alluvial canal. A canal flowing through those sediments transports some of this material along the flowing water. These canals take supplies from rivers which always carry sediments rolling on the bed or held in suspension, which is passed on to the off taking canals. If the velocity in a canal is very high, the suspension particles are not deposited, but when velocity is very low, the sediment held in suspension will get deposited.

Non Alluvial Canal: Non alluvial canals are canals that have been lined with some suitable material to provide a rigid bed banks so as to avoid the conflicts with the alluvial sides of a canal.

Inundations Canal: Those canals which depend for their supply on the periodical rise in canals are not always of the desired level. The water level of the river from which they are taken off. The supplies of these canals are not always of the assumed level. These canals are fill with water in Rainy season or in monsoon.

Permanent Canal: A canal is said to be a permanent canal when its source of supply is sufficiently well assured so as to warrant the construction of a regular grade channel supplied for the regulation and distribution. These canals are provided with a permanent masonry head works, distribution and regulator works. They are also constructed with engineering skills. The lining of the irrigation canal is also protective work because it helps in minimizing the chances of water- logging.

Feeder or Link Canal: Link canals supply water from a reservoir to another place wherefrom a given irrigation canal system is fed. These canals are used for diverting surplus water from one source to another.

Productive Canals: The canals, which indicate at the time of design and planning, that the total income will exceed the annual maintenance charges, are called productive canals.

Types of Canals

Watershed canal or Ridge canal: The dividing ridge line between the catchment areas of two streams is called the watershed or ridge canal.

Thus between two major streams, there is the main watershed, which divides the drainage area of the two streams.

The canal which is aligned along any natural watershed is called a watershed canal, or a ridge canal. Aligning a canal on the ridge ensures gravity irrigation on both sides of the canal.

Since the drainage flows away from the ridge, drainage cannot cross a canal aligned on the ridge. Thus, canal aligned on the watershed saves cost of construction of cross-drainage works.

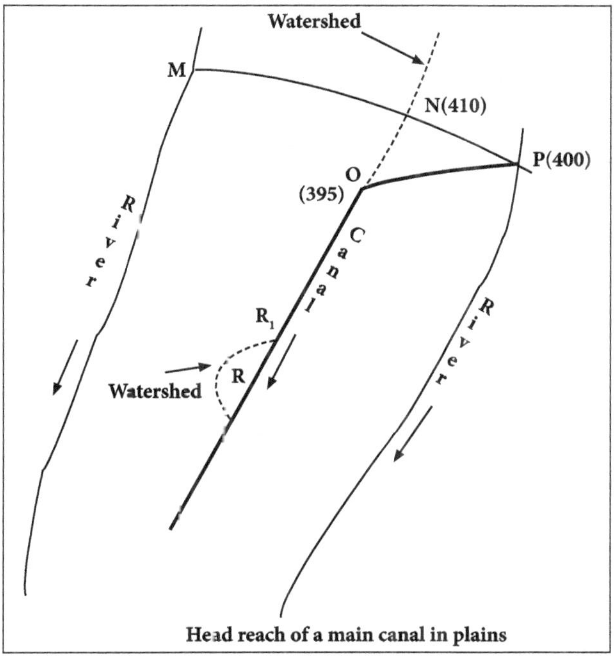

Ridge canal.

Contour Canal: Watershed canal along the ridge line are not found economical in hill areas. In hills, the river flows in the valley well below the watershed.

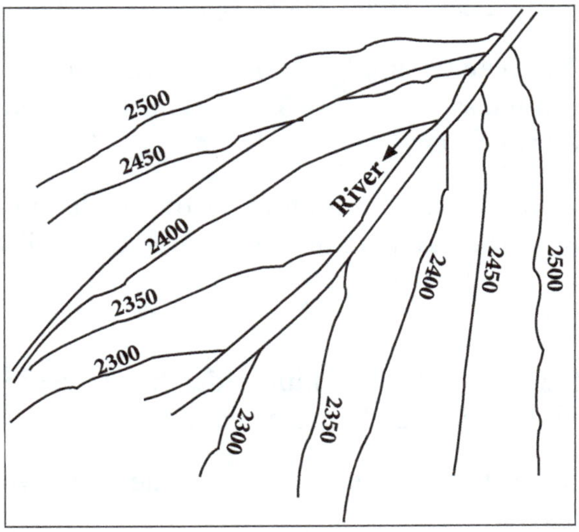

Contour canal.

In fact, the ridge line may be hundreds of meters above the river. It therefore becomes virtually impossible to take the canal on top of such a higher ridge line. In such conditions, contour canals are usually constructed.

A contour canal irrigates only one side because the area on the other side is higher.

Side Slope Canal: A side slope canal is one which is aligned at right angles to the contours; i.e. along the side slopes.

Since such a canal runs parallel to the natural drainage flow, it usually does not intercept drainage channels, thus avoiding the construction of cross-drainage structures. It is a canal which is aligned roughly at right angle to contours of the country but not on watershed or valley.

Side slope.

Types of Canals (Based on Lining)

Unlined Canal:

Bed and banks made up of natural soil:

- Water velocities higher than 0.7 m/s are not tolerable.
- High seepage and conveyance water losses.
- Profuse growth of aquatic weeds reduces the flow.

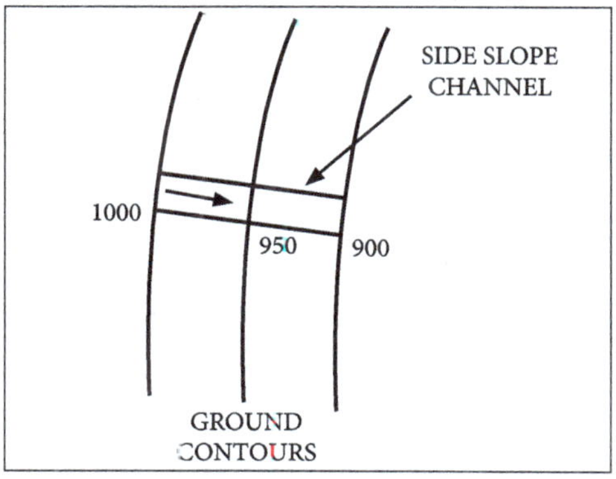

Unlined canal.

Lined Canal:

- Lining of impervious material on its bed and banks to prevent the seepage of water.
- Different types of lining used e.g. concrete, brick or burnt clay tile, boulder, etc.

Lined canal.

2.1.1 Design of Non-Erodible Channels

The initial dimensions of a channel are determined by uniform flow or manning formula. But final dimensions are calculated on the basis of:

- Hydraulic efficiency.

- Empirical rule of the best section.
- Practicability and Economy.

Factors Considered In the Design:

- Kind of material to find.
- Minimum Velocity.
- Maximum Velocity.
- Bed slopes.
- Side slopes.
- Free board.
- Kind of material is important to find roughness coefficient of channel.

Non erodible canal.

- Minimum Velocity is 2-3 ft/sec where it should non silting velocity to prevent aquatic growth.
- Maximum velocity is upto 8 ft/sec more than the above value, where the lining blocks are pulled away by moving water.
- Bed slope is dependent upon topography and energy required for flow of water.
- Side Slope: It Depends on the material forming the chemical section e.g. earth with lime stone h1: 1v.

Earth with concrete h1/2: 1v

- Free Board: Distance between top of channel to max water surface it should prevent waves it should be 5-13 % of depth.

Erodible Channels which Scour but do not silt. The behaviour of flow in erodible channels is influenced by several parameters and precise knowledge is not available on various aspects. Unlined channels with the channel bed and banks composed of earth, sand or gravel must be designed so that they maintain a stable configuration. There are three procedures:

- Velocity based Method of maximum permissible velocity.
- Regime Theory - Emprical equations for the channels with equilibrium sediment throughput.
- Shear Based - Tractive force methods, Shield analysis. The method of maximum permissible velocity also termed as non-erodible velocity: It is the highest mean velocity that will cause no erosion in the channel body.

When compared with the design process typically used for lined channels, the design of stable, unlined, erodible, earthen channels is a complex process involving numerous parameters, most of which cannot be correctly quantified. The complexity of the erodible channel design process results from the fact that in such channels stability is dependent not only on hydraulic parameters but also on the properties of material which composes of the bed and sides of the channel.

A stable channel section is one in which neither objectionable scour nor deposition occurs. There are three types of unstable sections: (USBR). The pioneering work of Fortier and Scobey in 1926 was the basis of channel design:

- The banks and bed of the channel are scoured but no deposition occurs. Example: When the channel conveys sediment free water but with adequate energy to erode the channel.
- Unstable channel with deposition but no scour. Example: When the water being conveyed carries a large sediment load at a velocity that permits sedimentation.
- Unstable channel with both scour and also deposition occur. Example: When the material through which the channel is excavated is susceptible to erosion and the water being conveyed carries a significant sediment load. These types of channels can be designed using the method of applying maximum permissible velocity.

The following important points are to be noted:

- First, the maximum permissible velocity is recommended for canals with a sinuous alignment.

- Second, these data are for depths of flow less than 0.91 m. For greater depths of flow, the maximum permissible velocity must be increased by 0.15 m/s.
- Third, the velocity of the in canals carrying abrasives, such as basalt raveling, should be reduced by 0.15 m /s.
- Fourth, channels diverting water from silt - laden river such as Ganga River should be designed for the mean design velocities 0.3 to 0.61 m/s greater than would be allowed for the same perimeter material if the water were transporting no sediment.

Following Steps are used for Designing provided a particular soil type, the channel is designed so that design velocity does not exceed Vmax for that soil and the channel side walls are with appropriate side slopes.

General guidelines: Froude number should be less than 0.35:

Step 1: For the given kind of material estimate roughness coefficient n, side slope m, and the maximum permissible velocity.

Step 2: Hydraulic mean radius is computed by using Manning formula.

Step 3: Area of flow is obtained using continuity equation.

Step 4: The wet perimeter is computed using the information obtained in steps 2 and 3.

Step 5: Solve simultaneously for b and y.

Step 6: Add proper free board and modify the section for practicality.

Example: A trapezoidal channel with bottom width of 6m, side slopes of 3H: 1V carries a flow of 50 m3 s-1 on a channel slope, So of 0.0015. The uniform depth flow for the channel is 1.3 m with n = 0.025. This channel is to be excavated in stiff clay. Check whether the channel will be susceptible to erosion or not.

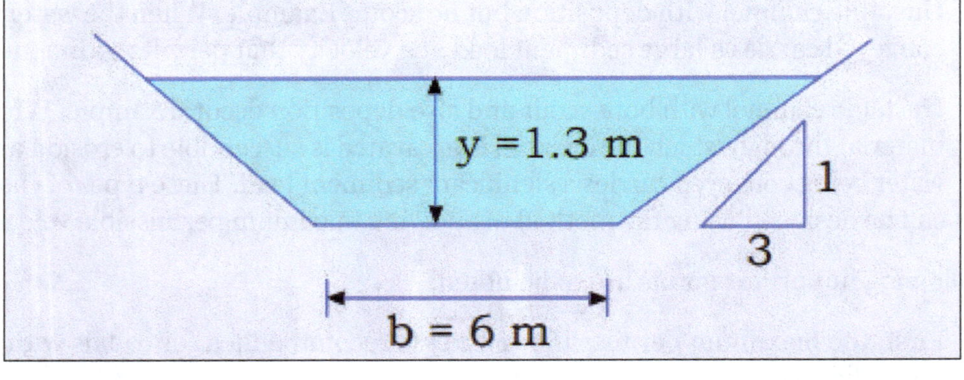

$$A = (b + my)y = (6 + 3 * 1.3) * 1.3 = 12.87 \text{ Sq.m}$$

$$\bar{v} = \frac{Q}{A} = \frac{20}{12.87} = 1.554 \text{ m s}^{-1}$$

Which is higher than the permissible velocity (of V = 1.25 ms^{-1}) from graph,

$$S_o = 0.0015 < 0.0065(0.65\%)$$

$$\therefore \text{Side slope adopted } 3:1 \text{ which is} < (1:1)$$

Increase width, b, to reduce velocity:

For b 8.4 m, y_n = 1.3 m Corresponding area of flow A = (8.4 + 3)*1.3=1599 m^2

$\bar{v} = \frac{Q}{A} = \frac{20}{15.99} = 1.251$ m/s which is equal to the permissible velocity.

2.2 Methods of Economic Section, Economics of Canal Lining and Design of Erodible Canals

Methods of Economic Section

If a canal section is such that the earth obtained from cutting can be fully utilized in forming banks, then that section is known as economical section. Again, the discharge will be maximum with minimum cross-section area. Here, no extra earth is needed from borrow pit and no earth is in excess to form spoil bank. This condition can arise only in case of partial cutting and banking. Sometimes, this condition is designated as balancing of cutting and banking. Here, the depth of cutting is called balancing depth.

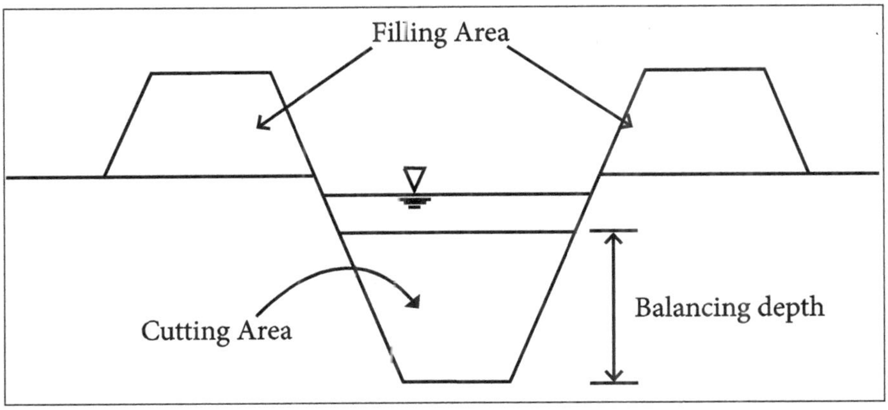

Economic Trapezoidal Channel

The continuity equation gives,

$$Q=A.V \quad ...(1)$$

Where A is the area of flow, V is the average velocity of flow.

The most widely used average velocity equation even till today is the Mannings equation, i.e.,

$$V=\frac{1}{n}\left(\frac{A}{p}\right)^{2/3} s_b^{1/2} \quad ...(2)$$

Where, P is the wetted perimeter of the canal and sh is the bed slope. n is called Meanings resistance coefficient, dependent on type of roughness of bed and sides. Hence. Equation (1) may be written as:

$$Q=A\cdot=\frac{1}{n}\left(\frac{A}{p}\right)^{2/3}(s_b)^{1/2}=\frac{1}{n}\frac{A^{5/3}}{p^{2/3}}s_b^{1/2} \quad ...(3)$$

or,

$$\frac{Qn}{A^{5/3}s_b^{1/2}}=\frac{1}{p^{2/3}}$$

$$\therefore p=\left[\frac{s_b^{3/4}}{(Qn)^{3/2}}\right]A^{5/2}=KA^{5/2} \quad ...(4)$$

Where,

$$K=\left[\frac{s_b^{3/4}}{(Qn)^{3/2}}\right]$$

For given discharge Q, Manning n and bed slope sh. %Vetted perimeter P is minimum when A is minimum. For minimum P and A, cost of excavation and lining used to prevent seepage and erosion will be minimum. Thus in the design of canal considering P to be minimum is termed as method of economic section. Further when Q is maximum, P is minimum (from Equation 10.3) for a given value of rt. sb and A. Again canals used for irrigation require high level at the point of delivery. Therefore, a small slope is necessary or slope is taken from the general topography of the land. If the lands arc undulating or uneven, canal falls are provided to maintain same small slope. Irrigation canals are mostly trapezoidal in shape. In lined canals side slope is taken 111: IV or I H: 1 V. Thus, the method 2 requires only to find best hydraulic section. i.e. Relationship of depth y with bottom width B, for efficient or economic section when P is minimum for maximum Q.

The Best Hydraulic or Economic Section

Since irrigation canals are mostly trapezoidal, discussion will be confined to trapezoidal channel only. For the trapezoidal section shown in Figure below,

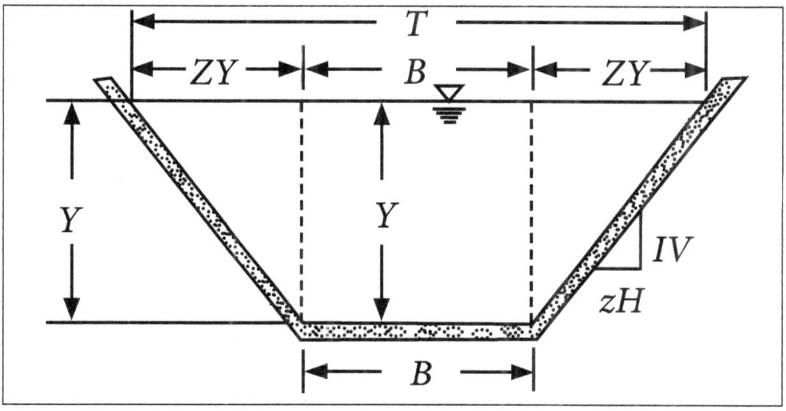

Trapeziodal section.

$$A = (B + zy)y = By + zy^2$$

$$\therefore \quad B = \frac{A}{y} - zy \qquad \text{...(5)}$$

Top width,

$$T = B + 2zy$$

$$p = B + 2y\sqrt{1+z^2}$$

or, $$p = \frac{A}{y} - zy + 2y\sqrt{1+z^2}$$

Assuming A and z to be constant P is differentiated with respect to y and equated to zero for minimum,

i.e.,

$$\frac{dp}{dy} = \frac{A}{y^2} - z + 2\sqrt{1+z^2} = 0$$

$$\therefore \quad \frac{A}{y^2} + z = 2\sqrt{1+z^2}$$

or,

$$\frac{By+zy^2}{y^2}+z=2\sqrt{1+z^2}$$

or,

$$\frac{B+zy}{y}+z=2\sqrt{1+z^2}$$

or,

$$B+zy+zy=2y\sqrt{1+z^2}$$

$$\therefore \left(\frac{B+2zy}{2}\right)=y\sqrt{1+z^2} \qquad ....(6)$$

i.e., half of top width = length of one slanting side, which is the condition for economic section. For the economic section, hydraulic radius (R AIP) is obtained as,

$$R=\frac{A}{p}=\frac{(B+zy)y}{B+zy\sqrt{1+z^2}}$$

Substituting the value of B from Equation (6) and simplifying,

$$R=\frac{y}{p} \qquad ....(7)$$

Thus, channel section is economic when hydraulic radius is half of the depth. Further if O is the centre of top with 7'. A perpendicular OA is drawn on side CD, B is the angle made by sloping side with horizontal, then from right angled triangle,

$$OAC, OA=OC\sin\theta=\frac{T}{2}\sin\theta$$

$$\therefore OA=\left(\frac{B+2zy}{2}\right)\sin\theta$$

And,

$$\sin\theta=\frac{y}{CD}=\frac{y}{y\sqrt{1+z^2}}=\frac{1}{\sqrt{1+z^2}}$$

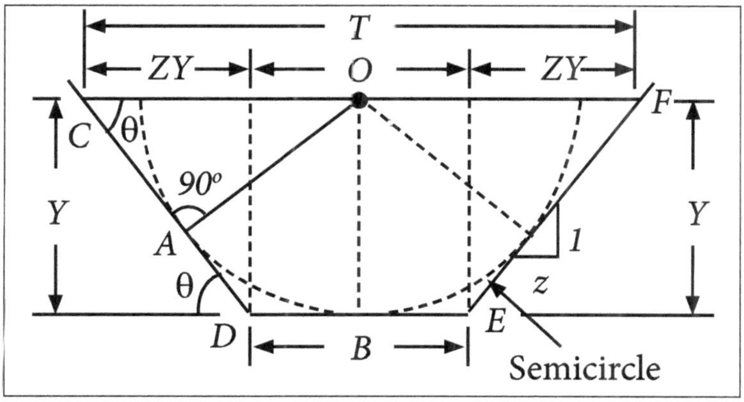

Economic section.

$$\therefore OA = \left(\frac{B+2zy}{2}\right)\frac{1}{\sqrt{1+z^2}}$$

$$\therefore OA = \sqrt{1+z^2} = \left(\frac{B+2zy}{2}\right) \qquad ...(8)$$

Comparing Equation (6) and (8) OA = y = depth This shows another condition, i.e., if a semicircle is drawn with O is centre and radius equal to depth of flow y. the three sides of economic trapezoidal channel will be tangential to this semicircle.

Economic Rectangular Channel

Consider below figure,

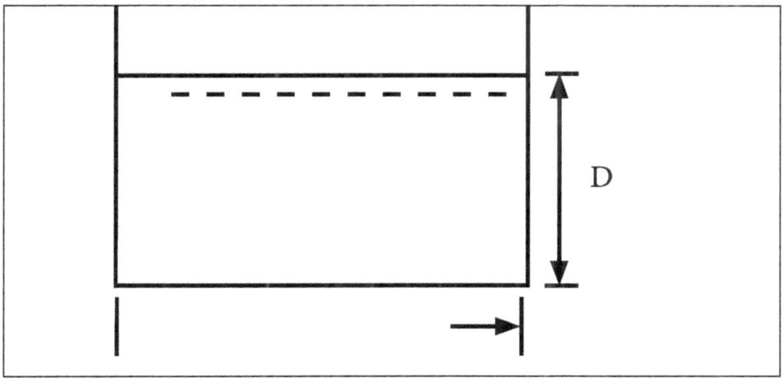

Rectangular section.

Let,

B = width of channel,

D = depth of flow.

$\therefore$ Area of flow, A = B x D, ...(1)

Wetted perimeter, P = 2D + B, ...(2)

From Equation (1) we have which if substituted in (2).

We get,

$$p = 2D + \frac{A}{D} \quad ...(3)$$

For most economical cross section, P should be minimum for a given area,

$$\frac{dP}{dD} = 0, \quad \text{so} \frac{dP}{dD} = 2 - \frac{A}{D^2} = 0 \Rightarrow 2 = \frac{A}{D^2} = \frac{BD}{D^2}$$

$$\Rightarrow \quad 2 = \frac{B}{D}$$

And hence,

$$D = \frac{B}{2} \quad ...(4)$$

The corresponding hydraulic radius is,

$$R_h = \frac{A}{p} = \frac{B \times D}{B + 2D} = \frac{2D \times D}{2D + 2D} = \frac{2D^2}{4D}$$

And hence,

$$R_h = \frac{D}{2} \quad ...(5)$$

From equations (4) and (5), it is clear that rectangular channel will be most economical when either: (a) the depth of the flow is half the width (Eq. 4), or (b) the hydraulic radius is half the depth of flow (eq. 5).

Most Economical Circular Channel

For a most economical section the discharge, for a constant cross-sectional area, slope of bed and resistance coefficient, is maximum (or P is minimum). But in the case of circular channels, the area of the flow cannot be maintained constant. Indeed, the cross-sectional area A and the wetted perimeter P both do not depend on D but they depend on the angle α. Referring to the figure shown, we can determine the wetted perimeter P and the area of flow A as follows:

Let,

D = depth of flow.

d = diameter of pipe.

r = radius of pipe.

2α = angle subtended by the free surface at the center (in radians).

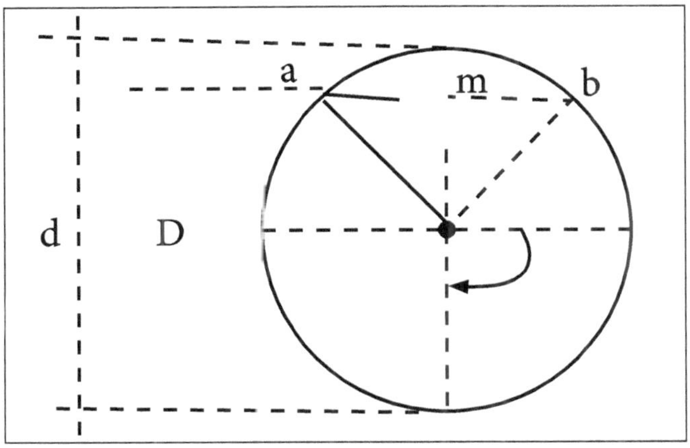

Circular section.

$$p = 2\alpha r = \alpha d \quad \text{...(A)}$$

$$A = A_1 + A_2$$

$$A_1 = \frac{1}{2}(2\alpha)r^2 = \alpha r^2 = \frac{\alpha r^2}{4}$$

$$A_2 = \text{area of } \Delta o \alpha b = \frac{1}{2} ab * OM = \frac{1}{2}(2 * Mb) * OM = Mb * OM$$

But,

$$Mb = \frac{d}{2}\sin\theta = \frac{d}{2}\sin(180° - \alpha) = \frac{d}{2}\sin\alpha$$

And,

$$OM = \frac{d}{2}\cos\theta = \frac{d}{2}\cos(180° - \alpha) = -\frac{d}{2}\cos\alpha$$

Therefore,

$$A_2 = -\frac{d^2}{4}\sin\alpha\cos\alpha = -\frac{d^2}{8}\sin 2\alpha,$$

And hence,

$$A = -\frac{ad^2}{4} - \frac{d^2}{8}\sin 2\alpha \quad \text{...(B)}$$

Consequently, the cross-sectional area A and the wetted perimeter P both depend on the angle α which is the most suitable variable. Thus in case of circular channels, for most economical section, two separate conditions are obtained:

- Condition for maximum discharge.
- Condition for maximum velocity.

Condition for Maximum Discharge for Circular Section: Using the Chezy formula,

$$Q = AV = AC\sqrt{R_h S} = C\sqrt{\frac{A^3}{p}s}$$

Squaring both sides of the equation.

We obtain,

$$Q^2 = C^2 s\left(\frac{A^3}{p}\right)$$

Recalling that A and P are both functions of α and applying the condition for maximum discharge,

$$\frac{dQ}{d\alpha} = 0.$$

We write,

$$2Q = \frac{dQ}{d\alpha} = C^2 S\left[3\frac{A^2}{p}\frac{dA}{d\alpha} - A^3\frac{1}{p^2}\frac{dP}{d\alpha}\right]$$

Simplifying and applying the above condition.

We can write,

$$3p\frac{dA}{d\alpha} - A\frac{dP}{d\alpha} = 0 \qquad \text{...(C)}$$

Now, let use find the derivatives involved,

$$\text{yields } \frac{dP}{d\alpha} = d \qquad \text{...(D)}$$

$$\text{yields } \frac{dA}{d\alpha} = \frac{d^2}{4} - \frac{d^2}{8}2\cos 2\alpha \qquad \text{...(E)}$$

Substituting Equation (A), (B), (D), and (E) into equation (C), we may write $3\alpha d\left[\frac{d^2}{4} - \frac{d^2}{4}\cos 2\alpha\right] - \left[\frac{\alpha d^2}{4} - \frac{d^2}{8}\sin 2\alpha\right] * d = 0$, after simplifying we get,

$$\frac{1}{2}\alpha d^3 - \frac{3}{4}\alpha d^3 \cos 2\alpha + \frac{d^3}{8}\sin 2\alpha = 0$$

or,

$$4\alpha - 6\alpha \cos 2\alpha + \sin 2\alpha = 0 \qquad ...(F)$$

The solution of this equation, by trial and error or graphically is,

$\alpha = 2\ 68.$ Radians

or,

$$\alpha = 154^\circ \qquad ...(G)$$

To find the corresponding depth of flow.

We write,

$$D = \frac{d}{2} + OM = \frac{d}{2} + \frac{d}{2}\cos\theta = \frac{d}{2} - \frac{d}{2}\cos\alpha = \frac{d}{2}(1 - \cos\alpha)$$

Substituting $\alpha = 154^\circ$

We obtain,

$$D = 0\ 95d \qquad ...(H)$$

This means that the maximum discharge (minimum P) in a circular channel occurs when the depth of flow is 0.95 times the diameter of the pipe. The above results holds good when the Chezy formula is used. If Manning's formula is used, results will be:

$$\alpha = 151^\circ \qquad ...(I)$$

And,

$$D = 0\ 94d \qquad ...(J)$$

Condition for Maximum Velocity for Circular Section

Since the cross-sectional area A varies with α, the condition for the maximum velocity is different from the condition for the maximum discharge. The condition for the maximum velocity may be obtained as follows:

Using the Chezy formula,

$$V = C\sqrt{R_h S} = C\sqrt{\frac{A}{p}S}$$

Squaring both sides of the equation.

We obtain,

$$V^2 = C^2 S\left(\frac{A}{p}\right)$$

Recalling that A and P are both functions of α and applying the condition for maximum velocity dV/ dα = 0.

We write,

$$D = \frac{d}{2} + OM = \frac{d}{2}(1 - \cos\alpha)$$

Simplifying and applying the above condition, we can write,

$$p\frac{dA}{d\alpha} - A\frac{dP}{d\alpha} = 0 \qquad \text{...(K)}$$

Substituting Equation (A), (B), (D), and (E) into equation (K), we may write,

$$\alpha d\left[\frac{d^2}{4} - \frac{d^2}{4}\cos 2\alpha\right] - \left[\frac{\alpha d^2}{4} - \frac{d^2}{8}\sin 2\alpha\right] * d = 0$$

After simplifying we get,

$$-\frac{1}{4}\alpha d^3 \cos 2\alpha + \frac{d^3}{8}\sin 2\alpha = 0$$

$$2\alpha \cos 2\alpha = \sin 2\alpha$$

$$\tan 2\alpha = 2\alpha \qquad \text{...(L)}$$

The solution of this equation, by trial and error or graphically is,

α = 2 25. radians or α = 128.750 ...(M)

To find the corresponding depth of flow, we have,

$$D = \frac{d}{2} + OM = \frac{d}{2}(1 - \cos\alpha)$$

Substituting α = 128.75°.

We obtain,

D = 0 81d ... (N)

Thus in the most economical circular section, the maximum velocity occurs when the depth of flow is 0.81 times the diameter of the pipe.

See that the maximum velocity occurs at the same depth even if Manning's formula was used.

Permissible Velocities in Channels

Higher velocities can be safely used in lined channels. Though maximum permissible velocities for concrete linings have not been established, velocities up to 2.5 m/sec. are permitted when the lining is not reinforced. The concrete lined Nangal Hydel Channel having capacity of 355 cumecs is designed for permissible average velocity of 1.8 m/sec. If still higher velocities are required, the lining can be reinforced accordingly. Asphaltic concrete, which has less resistance to abrasion, can withstand maximum velocity up to 1.5 m/sec. (5 ft./sec.). Values of limiting velocities in different kinds of linings, as per Indian Standards.

Table: Maximum Permissible Velocities in different types of Linings:

Type of lining	Permissible Velocities
Cement concrete lining	2.0 to 2.5 m/sec.
Burnt clay tile lining	1.8 m/sec.
Boulder lining	1.5 m/sec.

Dimensions of the section for erodible channel or unlined channel such as earth canal can be easily computed by the method of permissible velocity. The maximum permissible velocity or the non-erodible velocity is the greatest mean velocity that will not cause erosion of the channel body. An old and well-seasoned channel will stand much higher velocities than new ones. When other conditions are same, a deeper channel will convey water at a higher mean velocity without erosion than a shallower one.

Maximum permissible velocities entirely depend on the material that is used and the channel bed slope.

For example: in case of chutes, spillways the velocity may reach as high as 25 m/s. As the dam heights are increasing the expected velocities of the flows are also increasing which can reach as high as 70 m/s in exceptional cases. Thus, when one refers to maximum permissible velocity, it is for the normal canals that are built for irrigation purposes and Power canals in which the energy loss must be minimized. Hence, following table gives the maximum permissible velocity for some selected materials.

2.2.1 Economics of Canal Lining

The economic viability of lining of a canal is decided on the basis of the ratio of additional benefits derived from the lining to additional cost incurred on account of lining. The ratio is worked out as follows:

Let C = cost of lining in Rs./sq. metre including the additional cost of dressing the banks for lining and accounting for the saving, if any, resulting from the smaller cross-sections and, hence, smaller area of land, quantity of earth work, and structures required for the lined sections. This saving will be available on new canals excavated to have lined cross-section right from the beginning, but not on lining of the existing unlined canals.

s and S = seepage losses in unlined and lined canals, respectively, in cubic metres per square metre of wetted surface per day of 24 hrs.

p and P = wetted perimeter in metres of unlined and lined sections, respectively.

T = total perimeter of lining in metres.

d = number of running days of canal per year.

W = value of water saved in rupees per cubic metre.

L = length of the canal in metres.

y = life of the canal in years.

M = annual saving in rupees in operation and maintenance due to lining, taking into account the maintenance expenses on lining itself.

B = annual estimated value in rupees of other benefits for the length of canal under consideration. These will include prevention of waterlogging, reduced cost of drainage for adjoining lands, reduced risk of breach, and so on.

The annual value of water lost by seepage from the unlined section,

= pLsdW rupees.

The annual saving in value of water otherwise lost by seepage,

= (pLsdW – PL SdW) rupees.

= {LdW (ps – PS)} rupees.

Total annual benefits resulting from the lining of canal,

Bt = {LdW (ps – PS) + B + M} rupees.

Additional capital expenditure on construction of lined canal = TLC rupees

If the prevalent rate of interest is x per cent per year, the annual installment a (rupees) required to be deposited each year (at its beginning) for a number of y years to amount to TLC plus its interest at the end of y years is determined by the following equation,

$$TLC\left(1+\frac{x}{100}\right)^{y}=a\left\{\left(1+\frac{x}{100}\right)^{y}+\left(1+\frac{x}{100}\right)^{y-1}+\left(1+\frac{x}{100}\right)^{y-2}+\ldots+\left(1+\frac{x}{100}\right)^{1}\right\}$$

$$=\frac{a\left(1+\frac{x}{100}\right)\left\{\left(1+\frac{x}{100}\right)^{y}-1\right\}}{\left\{\left(1+\frac{x}{100}\right)-1\right\}}$$

$$\therefore a=\frac{(TLC)\left(\frac{x}{100}\right)\left(\frac{100+x}{100}\right)^{y-1}}{\left(\frac{100+x}{100}\right)^{y}-1}$$

For lining to be economically feasible, the value of a should be less than the annual benefit B_t i.e., the ratio B_t/a should be greater than unity.

If b/y, z are specified the equation can be solved explicitly for y and b. The cost of materials used in lining a channel can be specified in terms of the value of material used. This may be expressed as,

$$AR^{2/3}=\frac{\left(by+my^{2}\right)^{5/3}}{\left[b+2y\sqrt{1+m^{2}}\right]^{2/3}}-\frac{Q_{n}}{\sqrt{S_{O}}}$$

Solve the above equation for y,

$$y=\frac{\left[b/y+2\sqrt{1+m^{2}}\right]^{1/4}}{\left(b/y+m\right)^{5/8}}\left(Q_{n}/\sqrt{S_{O}}\right)^{3/8}$$

$$\text{Cost of bed material } C_b=\mu_B t_b\left(b+2b'\right)=Bb+k \text{ per unit length}$$

$$\text{Cost of side material } C_s=\mu_s t_s\left(2E+E'\right)=2\Gamma\left[\left(y+F_B\right)\sqrt{1+m^{2}}\right]$$

Therefore,

$$C=C_b+C_S=bB+k+2\Gamma\left[\left(y+F_B\right)\sqrt{1+m^{2}}\right]$$

Where,

C = total material cost per unit length.

C_b = material cost per channel base per unit length.

C_s = material cost of sides per unit channel length.

b' = bottom corner width.

t_b = thickness of the base material.

t = channel side lining thickness.

μ_B= cost of base lining material per unit volume.

μ_s= cost of side lining material per unit volume.

B = base material cost for specified thickness per unit area.

Γ = cost of side lining material for specified thickness per unit area.

F_B = vertical free board requirement.

E = wetted length of the side.

E' = side length of the free board.

2.2.2 Design of Erodible Canals

The lined canals are not designed by the use of Lacey's and Kennedy's theory, because the section of the canal is rigid. Manning's equation is useful for designing.

The design considerations are:

- The section should be economical.
- The velocity should be maximum so that the cross-sectional area becomes minimum.
- The capacity of lined section is not reduced by silting.
- Design Parameters for Circular Section:
- The bed is circular with its center at the full supply level and radius equal to full supply depth 'D'. The sides are tangential to the curve. However, the side slope is generally taken as 1:1.

Trapezoidal Section

The horizontal bed is joined with side slope by a curve of radius equal to full supply depth D. The side slope is generally kept as 1:1.

Lined channels are built for five primary reasons:

- To permit the transmission of water at high velocities through areas of deep or difficult excavation in a cost - effective fashion.
- To permit the transmission of water at high velocity at a reduced construction cost.
- To decrease canal seepage, thus conserving water and reducing the waterlogging of lands adjacent to the canal.
- To reduce the annual cost of operation and maintenance.
- To ensure the stability of the channel section.

The design of lined channels from the view point of hydraulic engineering is a rather elementary process which generally consists of proportioning an assumed channel cross section. Details of some typical cross section of lined channels used on irrigation projects in the India are given elsewhere. A recommended procedure for proportioning a lined section is summarized in table given below. In this table, it is assumed that the design flow Q D, the longitudinal slope of the channel So, the type of channel cross section e.g., trapezoidal, and the lining material have all been selected prior to the initiation of the channel design process.

Step	Process
1	Estimate n or C for the specified lining material and **So**
2	Compute the value of section factor $AR^{2/3} = nQ/\sqrt{S_O}$ or $AR^{1/2} = Q/\left(C\sqrt{S_O}\right)$
3	Solve section factor equation for **yn** given appropriate expressions for A and R
4	If hydraulically efficient section is required, then the standard geometric characteristics are used and **yn** is to be computed.
5	Check for Minimum permissible velocity if water carries silt and for vegetation. Froude number (Check Froude number and other velocity constraints such as. $\leq 0.8\ V \leq 5.5$ m/s Generally, Froude number should be as small as possible for Irrigation canals and should be less than 0.35. Higher Froude numbers is permitted as in the case of super critical flows such as in chutes, flumes. Decide the dimensions based on practicability.
6	Estimate Required height of lining above water surface. Required freeboard, Figure. Balance excavations costs, costs of channel lining and assess the needs to modify "Hydraulically efficient section".
7	Summarize the results with dimensioned sketch.

2.3 Kennedy's Silt Theory and Balancing Depth of Cutting

Kennedy's Silt Theory

Robert Kennedy was an engineer (1895) in Punjab Irrigation Department. He made extensive study in 22 irrigation canals and their distributaries in upper Bari Doab area of then Punjab, India. From the data he collected from those old stable canals, presented the following non-silting and non-scouring velocity V_c initially in fps unit as,

$$V_C = 0.84\ y^{0.64}$$

Converting this original equation is MKS unit system,

i.e.,

$$V_C = 0.547159 y^{0.64}$$
$$V_C = 0.55 y^{0.64}$$

Where V, is the non-silting and non-scouring velocity in m/sec and y is the depth in metre. Kennedy claimed that sediment size plays an important role and so he V defined a critical velocity ratio $(CVR) = m_r = \frac{V}{V_c}$.

Where V is the actual velocity,

$$\therefore V = m_r V_c$$
$$\therefore V = m_r \times 0.55\ y^{0.64}$$

He found m_r › I (i.e., 1.1 to 1.2) for sediment coarser than upper Bari Doab and m_r < I (0.8 to 0.9) for sediment finer than his canals. Actual velocity V is given by Ganguillet and Kuiters equation, i.e.,

$$V = \left[\frac{\left(23 + \frac{0.00155}{S_b} + \frac{1}{n_r}\right)}{1 + \left(23 + \frac{0.00155}{S_b}\right)\frac{n_r}{\sqrt{R}}}\right]\sqrt{Rs_b} \qquad \ldots(1)$$

Where n_r is called Kutter's nr.

Kennedy claimed that due to generation of eddies in the canal bed and its rise to the surface, flowing water gets the silt supporting power. These eddies are generated due to fricion of flowing water with channel surface. Vertical component of such eddie try to cause the sediment to move up while weight of sediment acts downward and thus, the sediment is kept in suspension and silting is avoided.

Steps Required to Design Canal by Kennedy's Theory

The known values are discharge Q, bed slope sb (decided by topography), CVR m_r, and Kutter's nr. Trial and error method has to be adopted to determine the design values of y and B. Side slope 1 V is assumed (if not given). i.e.. z varies from 0.5 to 2 for hard to loose soil. It is based on angle of repose of earth.

Steps:

- Assume a trial value of depth y.
- Find the velocity by the equation $V = 0.55 m_r y^{0.64}$.
- Find flow area $A = \frac{Q}{V}$.
- Find B from A - (B + zy)y where A is known from step (3), z is assumed or given. y is assumed in step (1)
- Find $P = B + 2y\sqrt{1+z^2}$.
- Find $R = \frac{A}{P}$.
- Now find velocity V by Kona's equation, i.e., Equation (1).
- If the velocity obtained in step (2) and in step (7) are not almost equal. Assume second trial values of depth y and repeat steps (1) to (8).
- Repeat this process, until velocity in step (2) - velocity in step (7). The value of y at which the two velocities are almost same, is the required depth of flow.
- Find B when y is known.
- Assume a reasonable free board (FB). Usually 0.6 when Q < 10 m^3/sec and FB > 0.75 when Q > 10 m^3/sec.

This trial and error method is very tedious and assumption of first trial values ay is very difficult. The design of canal by Kennedy's theory can be designed from Garret's diagram which provide a graphical solution of Kennedy's equation and Cianguillet–Kutter's equation with interpolation.

Drawbacks of Kennedy's Theory

- In his equation $V_c = Ky^r$. The values of K and it vary with sediment size. K = 0.55 and n=0.64 are only applicable to Bari Doab canal sediment, hence, these values cannot safely be applied to canal of different soil and sediment sizes.

- Kennedy did not give any equation or method of measurement of m, (CVR).
- Kennedy equation shows that velocity is only a function of depth y. Shape of channel, width, roughness, slope are not at all considered in assuming the velocity.
- Assumption of first assumed value of depth y to initiate trial and error method is very difficult. Some approximate value based on discharge. Slope, etc. should have been given to save the computational time.
- His regime velocity. i.e. Non-silting and nonscouring velocity did not consider the sediment load as variable.
- Gunguillet–Kutter's equation to check the mean velocity of flow with sediment has limitation as the equation is for nonerodible canal.
- He did not give any equation of slope.
- Complex silt carrying phenomenon cannot be represented by only one factor like in (CVR).
- y/B ratio which is dependent on Q was not noticed by Kennedy.
- It is entirely empirical and also requires rigorous trial and error which is tedious and time consuming.

2.3.1 Lacey's Theory: Concept, Equations and Limitations

Concept

Taking lead from Kennedy theory Mr. Gerald Lacey undertook detailed study to evolve more scientific method of designing the irrigation channels on alluvial soils. He presented revised version of his study in 1939 which is popularly known as Lacey's theory.

In this theory, Lacey explained in detail concept of regime conditions and rugosity coefficient.

It may be seen that for a channel to achieve regime condition following three conditions have to be fulfilled:

- Channel should flow uniformly in "incoherent unlimited alluvium" of same character as that transported by the water.
- Silt grade and silt charge should be constant.
- Discharge should be constant.

These conditions are very rarely achieved and are very difficult to maintain in practice.

According to Lacey's conception regime conditions may be subdivided as initial and final.

In rivers achievement of initial or final regime is practically impossible. Only in bank full stage or high floods the river can be considered to achieve temporary or quasi-regime. Recognition of this fact can be utilised to deal with the issues concerning scour and floods.

Lacey also a state that the silt is kept in suspension solely by force of eddies. Lacey adds that eddies are not generated on the bed only but at all points on the wet perimeter. The force of eddies may be taken normal to the sides.

Lacey's Theory

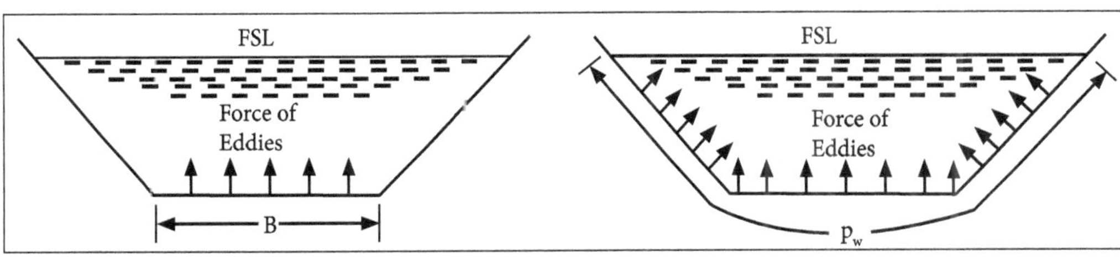

(a) According to Kennedy (b) According to Lacey

The vertical components of forces due to eddies are responsible for keeping the silt in suspension. Unlike Kennedy, Lacey takes hydraulic mean radius (R) as a variable rather than depth (D).

As wide channels are concerned there is hardly any difference between R and D. When the channel section is semi-circular there is no base width and sides actually.

Consequently amount of silt transported is not dependent on the base width of a channel only. On the basis of arguments Lacey plotted a graph between regime velocity (V) and hydraulic mean radius (R) and gave the relationship,

$$V = K.R1/ \qquad ...(1)$$

Where K is a constant.

Power of R is a fixed number and it doesn't need alteration to suit different conditions. Lacey recognized the importance of silt grade in the problem and introduced a concept of function 'f' known as silt factor.

He adjusted the values of such that it also came under square root sign. It gave scalar conception.

Equation (1) is thus modified as,

$$V = K. \sqrt{f}.R \qquad ...(2)$$

Kennedy's general equation is,

$$V = c.m.Dn \quad ...(3)$$

Comparing equations (2) and (3),

$$f = m2$$

From equation (2) it can be seen that in regime channels if the mean velocity is same then hydraulic mean radius varies inversely with the silt factor.

Lacey takes silt as a standard silt when silt factor is unity for that silt. He further states that standard silt is sandy silt within a regime channel with hydraulic mean radius equal to one metre.

Lacey's Regime Equations

After study and plotting of large data to justify this theory Lacey gave three fundamental equations from which other equations have been derived for design of irrigation channels.

The three fundamental equations are,

$$V = 0.639\sqrt{fR} \quad ...(1)$$

Where V is regime velocity in m/sec,

$$Af^2 = 141.2V^5 \quad ...(2)$$

$$V = 10.8R^{2/3}S^{1/3} \quad ...(3)$$

Where S is the slope of water surface.

Equation (3) is termed as the regime flow equation and is of great practical importance.

It may be seen that the equation does not contain the term of rugosity coefficient. It is necessary to know the value of coefficient of rugosity, the selection of which remains a matter of experience and many not be reliable especially in case of rivers in floods.

Considering that at high flood river flows in quasi-regime the regime flow equation (3) given above may be adopted though it may have some exceptions.

Important equations given by Lacey in his theory are summarised below.

The first three equation are called fundamental equations on the basis of which other equations have been developed.

$$V = 0.639\sqrt{fR} \quad \ldots 4 \qquad \ldots(4)$$

$$Af^2 = 141.24V^5 \quad \ldots 5 \qquad \ldots(5)$$

$$V = 10.8R^{2/3}S^{1/3} \quad \ldots 6 \qquad \ldots(6)$$

$$V = \frac{1}{n_\alpha}R^{3/4}S^{1/2} \quad \ldots 7 \qquad \ldots(7)$$

$$N_\alpha = 0.0225\,f^{1/4} \quad \ldots 8 \qquad \ldots(8)$$

$$V = 0.4382\left(Q.f^2\right)^{1/6} \quad \ldots 9 \qquad \ldots(9)$$

$$R = 2.46\frac{V^2}{f} \quad \ldots 10 \qquad \ldots(10)$$

$$p_w = 4.825\sqrt{Q} \quad \ldots 11 \qquad \ldots(11)$$

$$S = \frac{f^{5/3}}{3316Q^{1/6}} \quad \ldots 12 \qquad \ldots(12)$$

$$A = 2.28\frac{Q^{5/6}}{f^{1/3}} \quad \ldots 13 \qquad \ldots(13)$$

$$f = 1076\sqrt{m_r} \quad \ldots 14 \qquad \ldots(14)$$

$$R = 1.35\left(\frac{q^2}{f}\right)^{1/3} \quad \ldots \qquad \ldots(15)$$

Lacey's Regime Equations

Although all the above equations are given for regime conditions and are normally for canals in alluvium as the river attains quasi-regime conditions equations (6), (11) and (15) are very useful in calculating flood discharge, waterway needed during floods and scour depth during floods respectively.

When equation (15) is applied to rivers in flood, the value of R gives normal depth of scour. This formula is very useful determining the levels of foundations, vertical cutoffs etc. This formula is popularly called Lacey's scour depth formula.

Limitations of Lacey's Theory

- The Lacey's work is based on the field observations and empirically derived equations and therefore it cannot be said to be theory in the strict sense.
- Regime equations in their derived from cannot be applied universally. This is because it hold good mostly for the regions whose data was taken for study.
- Like Kennedy's theory even though perfect definition of silt grade and silt charge is not given in most of the equations are based on the silt factor 'f'.

- In practice regime condition stated by Lacey is rarely achieved and that too after a long period.
- The field observations show limited acceptance of the concept of semi-elliptical section of a regime channel.
- Complex phenomenon of sediment concentration and transport has not been scientifically considered.

Design of Irrigation Channels Making Use of Lacey Theory

The full supply discharge for any canal is always fixed before starting a design. The value of 'f' for particular site can be calculated using equation (11) or if CVR is given then f = m2.

Thus when Q and f are known design can be done in the following steps:

- Find out F using equation (6)

  V = 0.4382 (Q . f2)1/6.
- Calculate value of R using equation (7)

  R = 2.46 V2/f.
- Calculate wetted perimeter Pw using Lacey's regime perimeter equation Pw = 4.825 Q1/2.
- Calculate cross sectional area A from equation Q = AV.
- Assuming side slopes, calculate the full supply depth from A, Pw and R.
- Calculate the longitudinal slope of the channel using equation (9).

Example: Design of irrigation canal by Lacey's theory for the following data:

Solution:

FSD = 14 m^2/sec

f = 1

Side slopes = 1/2 : 1 (Horizontal : Vertical)

Coefficient of rugosity. N = 0.225

Solution: Mean velocity $V = 0.4382\,(14 \times 1^2)^{1/6} = 0.68$ m/sec.

Hydraulic mean radius,

$$R = 2.46\frac{(0.68)^2}{1} = 1.14\text{m}$$

Wetted perimeter,

$$P_w = 4.825 \times (14)^{1/2} = 18.06 \text{ m}$$

Cross-sectional area,

$$A = R \times P_w = 1.14 \times 18.06 = 20.55 \text{ m}^2$$

From Figure,

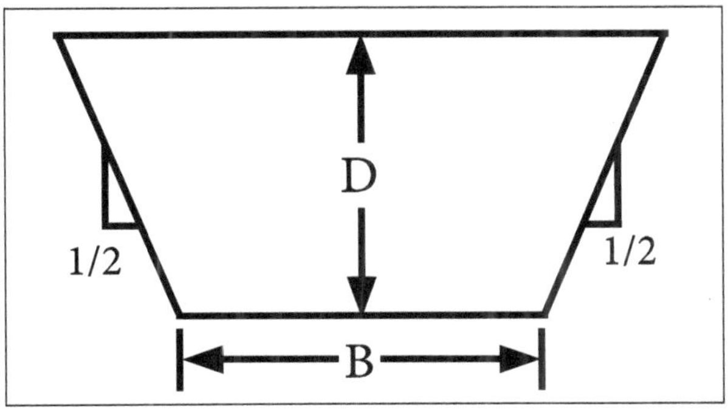

Problem 3.

$$A = BD + \frac{D^2}{2} \quad ...(1)$$

$$P_\omega = B + \sqrt{5}\ D \quad ...(2)$$

Putting value of A and P_w in equation (1) and (2) respectively,

$$20.55 = BD + D^2/2$$
$$18.06 = B + \sqrt{5}D \quad ...(3)$$

From equation (3),

$$B = 18.06 - \sqrt{5}\ D$$

Now putting Value of B in equation (1),

$$20.55 = \left(18.06 - \sqrt{5}D\right)D + D^2/2$$

or,

$$20.55 = 18.06D - 1.73\ D^2$$

or,

$$1.73D^2 - 18.06D + 20.55 = 0$$

It is a quadratic equation in D. Hence. Solving for D,

$$D = \frac{\sqrt{18.06} \pm \sqrt{(18.06)^2 - 4 \times 1.73 \times 20.55}}{2 \times 1.73}$$

$$D = \frac{18.06 \pm 13.56}{3.46}$$

or,

D = 9.13 and 1.3 m

Taking,

D = 9.13 m

B = 18.06 – 2.23 × 9.13 = – 2.14 which is absurd

Hence, taking

D = 1.3 m

B = 18.06 – 2.23 × 1.3 = 15.15 m

By using the value of B and D hydraulic mean radius can be calculated again to compare it with the value already calculated above using Lacey's equation:

$$R = 2.46 \frac{(0.68)^2}{1}$$

$$R = \frac{A}{P} = \frac{15.15 \times 1.3 + (1.3)^2 / 2}{15.15 + \sqrt{5} \times 1.3} = \frac{20.54}{18.05} = 1.14\,m$$

Now slope can be determined from formula,

$$S = \frac{f^{5/3}}{3316 \times Q^{1/6}} = \frac{1^{5/3}}{3316 \times (14)^{1/6}} = \frac{1}{3316 \times 1.553} = \frac{1}{5160}$$

2.3.2 Balancing Depth of Cutting

A canal section will be economical when earth work involved at a particular section has an equal amount of cut and fill. Usually a canal section has a part in cutting and part in filling as shown in figure. If the amount of cut is equal to the amount of fill, it has to be paid for once only.

For a given C/S there is always only one depth of cutting for which cutting and filling will be equal. The depth is known as balancing depth.

Balancing canal depth comes when the canal is in partially embankment and partially in cutting. It is the depth of the canal (H) which gives equal amount of filling and cutting. For a given cross-section of a canal, it has only one balancing depth. For this depth, canal sectional will be economical. B = Bed width.

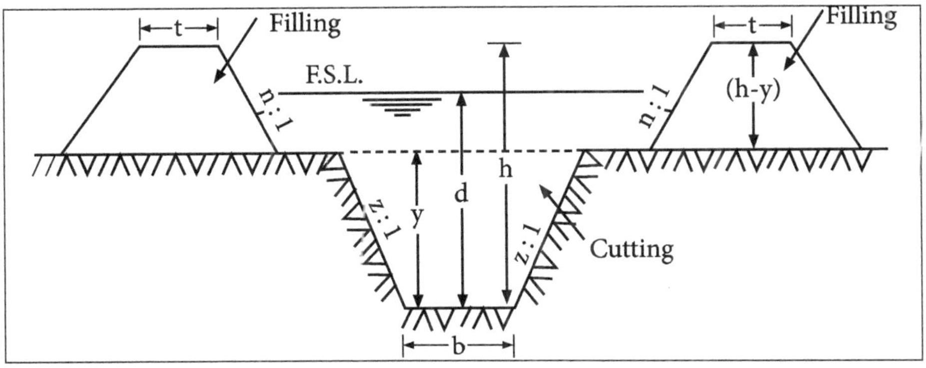

Balancing depth.

If,

h = vertical height of top of bank from the bed of canal.

b = bed width of the channel.

t = top width of the canal bank.

n:1 = side slope of bank in filling.

z:1 = side slope of the canal in cutting.

d = full supply depth of canal.

y = depth of cutting

Area of the cut = by + zy2 = y(b + zy)

Area of fill = 2[(h – y)t + n(h-y)2]

On equating the area of cut and fill,

y(b + z y) = 2[(h – y)t + n(h-y)2]

b y+ zy2 = 2th + 2nh2 – 2nhy – 2ty – 2nhy + 2ny2 y2 (2n – z) – (b + 4nh + 2t) y + 2h(t + nh) = 0

From this equation, balancing depth of the canal may be determined.

A canal is usually constructed with side slope of 1:1 in cutting and a slope 1.5:1 in filling.

Putting,

n = 1.5 and z =1 in above equation.

We get,

$$y2 - (b/2 + 3h + t)y + h (t + 3/2 h) = 0.$$

Chapter 3

Canal Structures

3.1 Falls: Types and Location

Irrigation canals are constructed with some permissible bed slopes so that there is no silting or scouring in the canal bed. But it is not always possible to run the canal at the desired bed slope throughout the alignment due to the fluctuating nature of country slope. Generally, the slope of the natural ground surface is not uniform throughout the alignment. Sometimes, the ground surface may be steep and sometimes it may be very irregular with abrupt change of grade. In those cases, a vertical drop is provided to step down the canal bed and then it is continued with permissible slope until another step down is necessary. This is done to avoid unnecessary huge earth work in filling. Such vertical drops are known as canal falls or simply falls.

Necessity of Canal Falls

When the slope of the ground suddenly changes to steeper slope, the permissible bed slope cannot be maintained. It requires excessive earthwork in filling to maintain the slope. In such a case falls are provided to avoid excessive earth work in filling.

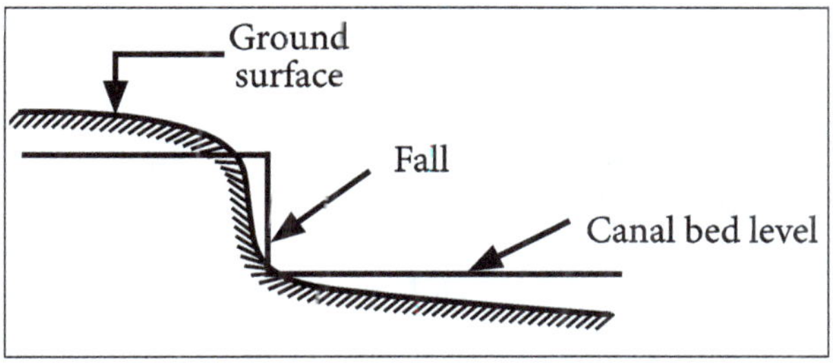

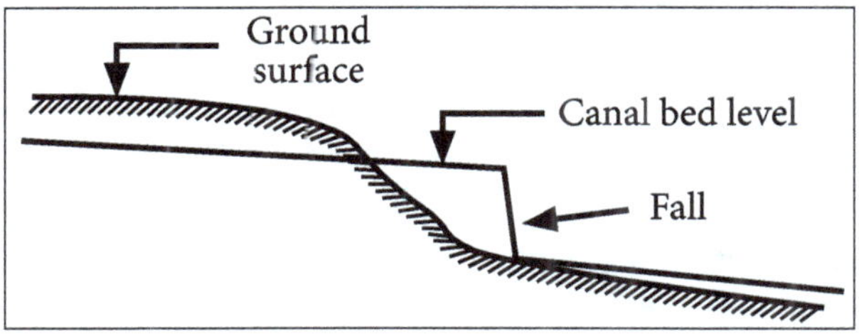

When the slope of the ground is approximately uniform and the slope is greater than the permissible bed slope of canal. In that case also the canal falls are needed.

In cross-drainage works, either when the difference between bed level of canal and that of drainage is small or when the F.S.L of the canal is above the bed level of drainage then the canal fall is necessary to carry the canal water below the stream or drainage.

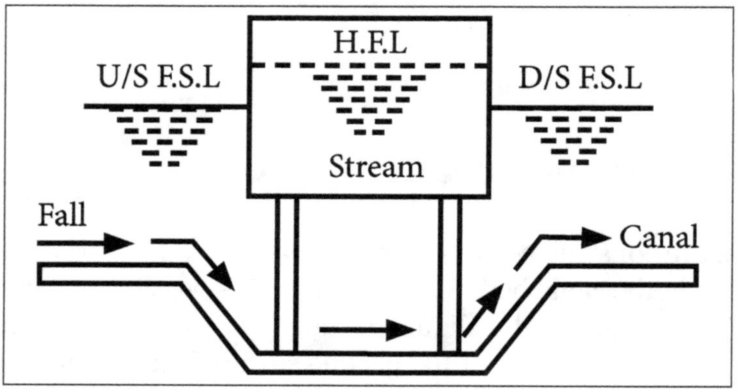

A canal fall is a hydraulic structure constructed across a canal to lower its water level. This is achieved by negotiating the change in bed elevation of the canal necessitated by the difference in ground slope and canal slope. The necessity of a fall arises because the available ground slope usually exceeds the designed bed slope of a canal. Thus, an irrigation channel which is in cutting in its head reach soon meets a condition when it has to be entirely in filling. An irrigation channel in embankment has the disadvantages of:

- Higher construction and maintenance cost.
- Higher seepage and percolation losses.
- Adjacent area being flooded due to any possible breach in the embankment.
- Difficulties in irrigation operations. Hence, an irrigation channel should not be located on high embankments.

Therefore, falls are introduced at appropriate places to lower the supply level of an irrigation channel. The canal water immediately downstream of the fall structure possesses excessive kinetic energy which, if not dissipated, may scour the bed and banks of the canal downstream of the fall. This would also endanger the safety of the fall structure. Therefore, a canal fall is always provided with measures to dissipate surplus energy which, obviously, is the consequence of constructing the fall. The location of a fall is primarily influenced by the topography of the area and the desirability of combining a fall with other masonry structures such as bridges, regulators, and so on. In case of main canals, economy in the cost of excavation is to be considered. Besides, the relative economy of providing a large number of smaller falls (achieving balanced earth work and ease in construction) compared to that of a smaller number of larger falls (resulting in reduced construction cost and increased power production) is also worked out. In case of channels which irrigate the command area directly, a fall should be provided before the bed of the channel comes into filling. The full supply level of a channel can be kept below the ground level for a distance of up to about 500 metres downstream of the fall as the command area in this reach can be irrigated by the channels off-taking from upstream of the fall.

Types of Canal Falls

The following are the different types of canal falls that can be adopted according to the site condition:

Ogee Fall

In this fall, an ogee curve is provided for carrying canal water from higher level to the lower level. This fall is suitable when the natural ground surface suddenly changes to a steeper slope along the alignment of the canal:

- The fall consists of the concrete vertical wall and the concrete bed.
- Over the concrete bed the rubble masonry is present in the shape of ogee curve.
- The surface of the masonry is finished with rich cement mortar.
- The upstream and downstream side of the fall is protected by stone pitching with cement grouting.
- The design considerations of the ogee fall depends on the site condition.

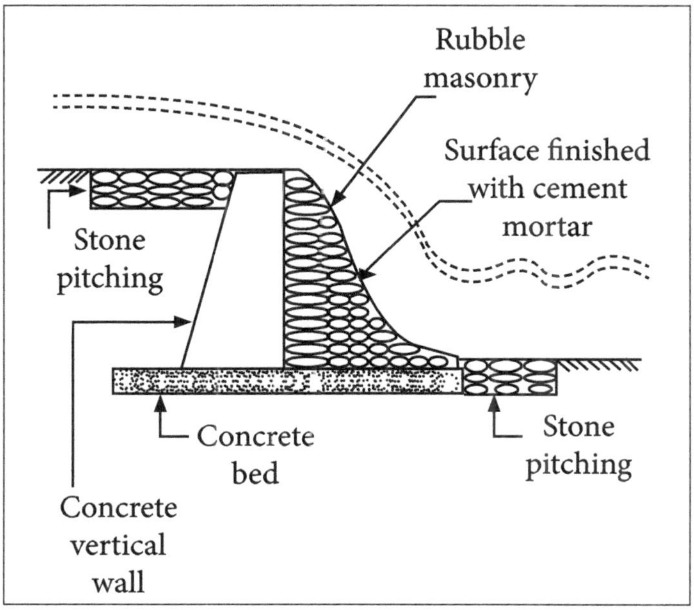

Ogee fall.

Rapid Fall

The rapid fall is apt when the slope of the natural ground surface is even and long. It has long sloping glacis with the longitudinal slope which varies from 1 in 10 to 1 in 20:

- Curtain walls were provided on the upstream and downstream side of the sloping glacis.
- The sloping bed is provided along with rubble masonry.
- The upstream and downstream side of the fall is also protected by rubble masonry.
- The masonry surface is finished with the rich cement mortar.

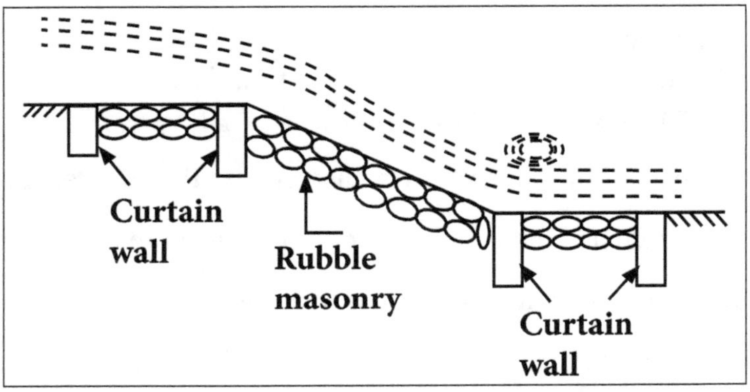

Rapid fall.

Stepped Fall

Stepped fall consists of series of vertical drops in the form of steps. This fall is suitable in places where sloping ground is really long and requires long glacis to connect the higher bed level with lower bed level:

- This fall is practically a modification of the rapid fall.
- The sloping glacis is divided into a number of drops so that the flowing water will not cause any damage to the canal bed. Brick walls are provided at each of the drops.
- The bed of the canal within the fall is protected by rubble masonry with surface finishing by rich cement mortar.

Trapezoidal Notch Fall

In this fall a body wall is constructed across the canal. The body wall consists of number of trapezoidal notches between side piers and intermediate pier. The sills of the notches are kept at the upstream bed level of the canal:

- The body wall is constructed with the concrete or masonry.

- An impervious floor is then provided to resist scoring effect of falling water.
- The upstream and downstream side of fall is protected by means of stone pitching finished by cement grouting.
- The size and number of notches depends upon the full supply of the canal discharge.

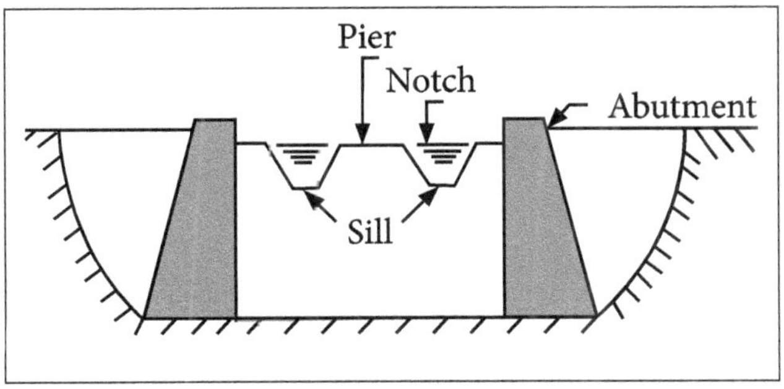

Trapezoida notch fall.

Vertical Drop Fall

It consists of a vertical drop walls constructed with masonry work. The water flows over the crest of the wall. A water eastern is provided on the downstream side that acts as a water cushion in order to dissipate energy when the water falls:

- A concrete floor is present on the downstream side to control the scouring effect of the flowing water.
- Curtain walls are present on the upstream and downstream side.
- Stone pitching with cement grouting is provided on the upstream and downstream side of the fall so as to protect it from scouring.

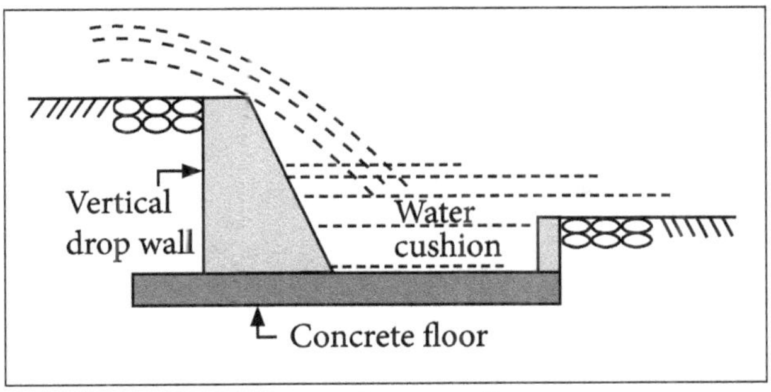

Vertical Drop Fall.

Glacis Fall

It consists of a straight sloping glacis provided along with the crest. A water cushion is provided on the downstream side to dissipate the energy of flowing water.

- The sloping glacis is constructed with cement concrete.
- Curtain walls and toe walls are provided on the upstream and the downstream side.
- The space between the toe walls and curtain walls is protected by means of stone pitching.
- This type of fall is suited for drops up to 1.5 m.

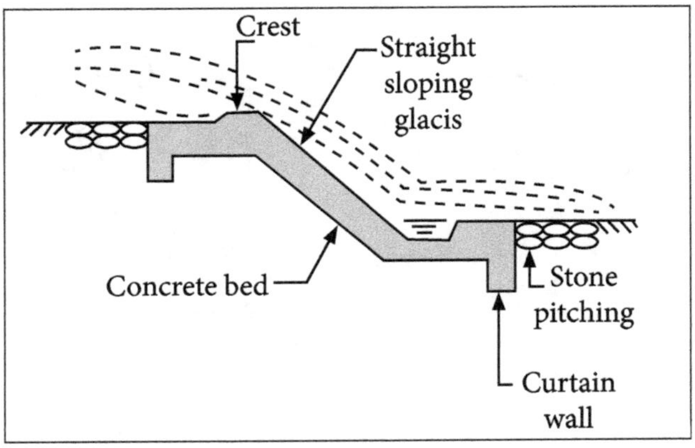

Glacis Fall.

1. Montague Type Fall

In this type of fall, the straight sloping glacis is modified by giving a parabolic shape which is termed as the Montague profile. Taking "o" as the origin, the Montague profile is given by the equation,

$$X = v\sqrt{\frac{4y}{g}} + Y$$

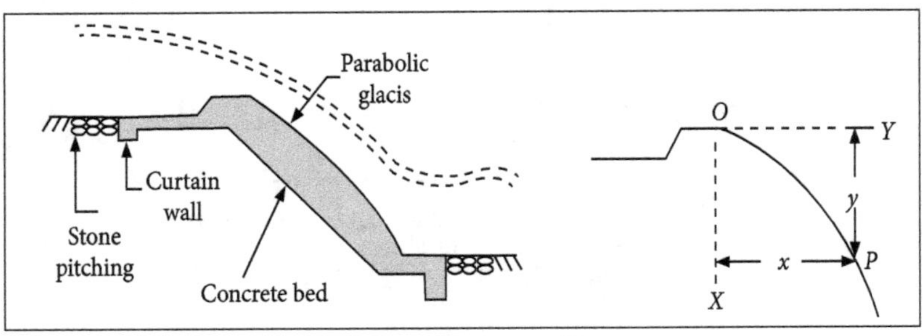

Montague Type Fall.

Where, x = distance of point P from OX axis, Y = distance of point P from OY axis, υ = velocity of water at the crest, g = acceleration due to gravity.

The main body of the fall is constructed with cement concrete. Toe walls, curtain walls and bed protection by stone pitching are same as in the case of straight sloping glacis.

2. Inglis Type Fall

In this type of fall, the gracis is straight and sloping, but buffle walls are provided on the downstream floor to dissipate the energy of flowing water.

- The height of buffle depends on the head of water on the upstream side.
- The main body of the fall is constructed with the cement concrete.
- The toe walls and curtain walls are same as that of straight glacis.
- The protection works with stone pitching are also same. Sometimes, this fall is called as buffle fall.

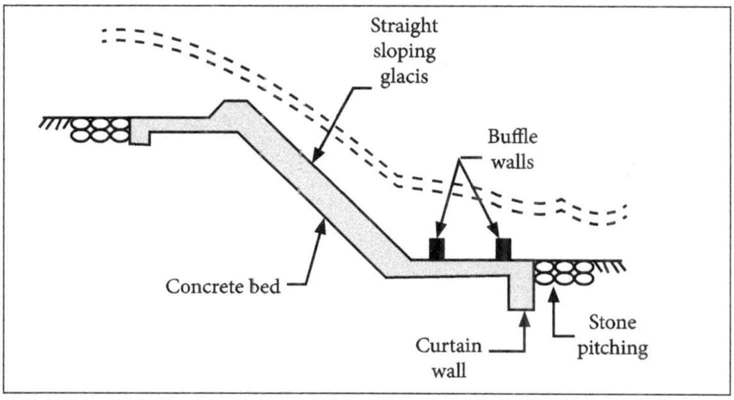

Inglis Type Fall.

3.1.1 Design Principles of Sarda Type Fall and Straight Glacis Fall

Such type of falls are constructed on Sarda canal in Uttar Pradesh. It is a fall with raised crest and with the vertical impact. The soils in Sarda command comprised sandy stratum overlain by sandy-clay on which the depth of cutting was to be kept minimum. This made it obligatory to provide the number of falls with small drops. In Sarda type falls (q) discharge intensity varied from 1.6 to 3.5 cumec/m and the drop varied from 0.6 to 2.5 m.

Crest Dimensions

This type of fall is not flumed.

For canal discharge 15 cumec,

Crest length of fall = Bed width of the canal.

For distributaries and minors,

Crest length of fall = Bed width + Depth of flow.

Body wall: When discharge of the canal is less than 14 m3/sec then the section of body wall is kept rectangular (Figure (a)).

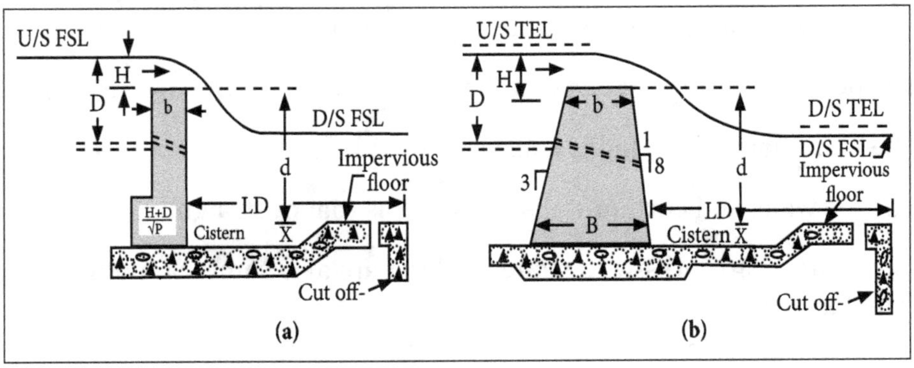

When the discharge of a canal is more than 14 m³/sec then the section of the body wall is kept trapezoidal with upstream batter 1: 3 and downstream batter 1: 8.

For Rectangular Body Wall

Base width 'B' = H + d/√p

Top width 'b' = 0.552 √d

For trapezoidal body wall Top width b = 0.522 √(H + d)

The edges are rounded with the radius of 0.3 m.

Base width B is calculated by the batter given to u/s and d/s sides.

Here H is depth of water above the crest of the fall in metres.

d is the height of the crest above the downstream bed level in metres.

Discharge Over Crest

Under free fall condition the discharge formula used in this type of fall is,

$$Q = CLH \{H/b\}1/6$$

Where L is length of crest in m and Q is discharge in cumec.

Value of C for trapezoidal crest is 2 and for rectangular crest 1.85.

For submerged flow conditions neglecting velocity of approach the discharge is given by the following formula:

$$Q=\frac{2}{3}Cd\cdot L\cdot\sqrt{2g}\ (H_L)^{3/2}+Cd\cdot L\cdot h_2\sqrt{2g\cdot H_L}$$

Where,

$Cd = 0.65$

HL = drop in water surface.

And,

h_2 = depth of d/s water level over top of crest.

Crest Level

The height of crest above the upstream bed level is fixed in a way that the depth of flow u/s of the fall is not affected. From the discharge formula mentioned above, as Q is known value of H can be calculated.

R . L of crest = F . S . L on the u/s – H.

The stability of body wall should be tested by the usual procedure when the drops exceeding 1.5 m are to be designed. In the body wall drain holes can be provided at the u/s bed level to dry out the canal during closures for maintenance, etc.

Cistern dimensions: Dimensions of the cistern can be fixed from the Bahadurabad Research Institute formula given, i.e.,

$LC = 5\sqrt{E.HL}$

$X = ¼ (E.HL)2/3$

Total Length of Impervious Floor

Since for any hydraulic structure total length of impervious floor must be designed on the basis of the Bligh's theory for small structures and Khosla's theory for someother works. The maximum seepage head is experienced when on u/s water is upto the crest level of the fall and also when there is no flow on the d/s side. Referring the Figure maximum seepage head is given by 'd'.

Length of d/s Impervious Floor

The maximum length of the d/s impervious floor is given by the following relation.

Ld =x 2D + 2.4 + HL in metres.

The balance of the impervious floor can be provided under body wall and on u/s.

Thickness of the Floor

The d/s floor must be made thick enough to resist the uplift pressures. However, minimum thickness of 0.3 to 0.6 m of concrete under 35 cm of brick masonry can be provided on the d/s. Masonry is not necessary on the u/s brick. The brick on the edge laid on the d/s impervious concrete floor provide the additional strength and affords easy repairs to the floor.

Cut-off

A required depth of cut-off below the floor must be provided at the d/s end of the floor so as to provide safety against steep exit gradient. The depth of cut-off can range from 1 to 1.5 m. Sometimes deeper cut-offs may be necessary to reduce horizontal floor length to satisfy Khosla's principle of exit gradient. For falls having 1 m and above head on the crest must be provided more cut-offs. Cut-off at u/s end of floor is provided and that may be smaller in depth.

Other Protective Works

Providing other accessories such as upstream wings, staggered blocks on the cistern floor, downstream wings, bed and side pitching is generally done on the basis of the thumb rules. For big structures, actual design calculations may be done. For general arrangement see Figure below:

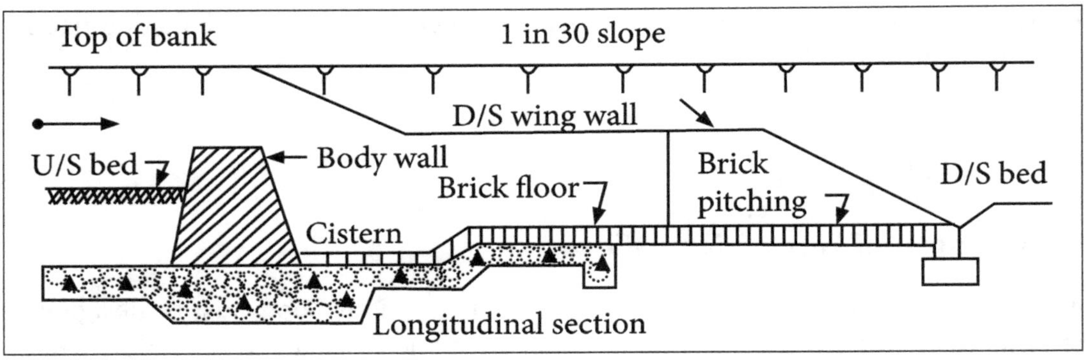

Upstream Wing Walls

For small falls upto 14 cumec upstream wings can be splayed at 1: 1. For higher discharges u/s wing walls are kept segmental with a radius equal to 6 H and continued thereafter tangentially merging to the banks. The wings cn also be embedded into the bank for about 1 m.

Downstream Wing Walls

For the length of the cistern d/s wing walls are kept vertical from the crest. Thereafter they are wasped or flared to a slope of 1: 1. An average splay of 1 in 3 for attaining required slope is given to the top of the wings. The wings can be taken deep into the banks.

Staggered Blocks

Staggered block of height dc must be provided at a distance of 1.0 dc to 1.5 dc from the d/s toe of crest for clear falls. In case of submerged falls the blocks may be provided at the end of the cistern. A row of staggered cubical blocks of height equal to 0.1 to 0.13 of depth of water is invariably provided at the end of the d/s impervious floor.

Bed and Side Pitching

The d/s bed pitching with bricks 20 cm thick over 10 cm ballast is provided horizontally for a length of 6 m. Thereafter for lengths up to 5 to 15 m and for falls varying from 0.75 to 1.5 m can be provided with down slope of 1 in 10. The side pitching with bricks on edge with 1: 1 slope is provided after the return-wing on the downstream. A toe wall should be provided between the bed pitching and the side pitching to provide a firm support to the latter.

Design Principles for Straight Glacis Fall

Crest Dimensions

Clear Width of Crest: Vertical falls must be full width falls, i.e., the width of the crest should be of similar bed width of the canal because increased intensity of the discharge due to fluming creates scour on the downstream.

Unlike vertical falls the glacis falls may be flumed when combined with bridge so as to economize in the cost. It k quite rational to select such (q) discharge per metre run of crest width with which the height of drop (HL) available gives value of total energy on the d/s (Ef2) equal to F.S. depth of the canal. It does not need deep cistern on d/s and avoids construction difficulty particularly when the subsoil water level is high. The throat width can be rounded off to next half metre. The fluming thus calculated may not exceed its limit given below .It is subjected to the condition that overall width of fall crest is not more than the bed width of the canal on the downstream.

Drop (H_L)	Permissible fluming ratio
Up to 1 m	66%
1 to 3 m	75%
Above 3 m	85%

Crest level = u/s TEL – E

In case of full width falls and sometimes in flumed falls if the crest level works out unreasonably then high fluming may be done or increased .If already flumed so that crest is not higher than 0.4 –D1,then above the u/s bed as otherwise it will increase afflux at low supplies and may cause alternate silting/scouring.

The value of E is calculated based on the discharge formula Q = 1.84 Bt X E3/2.

Where Bt is clear width of crest. Therefore if n piers are provided between effective,

Bt = (Bt – 0.2 n H)

And E is depth of crest below u/s TEL.

Length of crest (Lt) = 2/3 E.

The crest is joined to u/s and d/s canal bed with sloping glacis.

The u/s glacis (for non-meter falls) is given a slope of 1/2: 1. The u/s crest end is kept curved with a radius of E/2,

The d/s glacis is given 2: 1 slope and it is joined with the cistern d/s along the curve having radius equal to E.

Cistern Dimensions

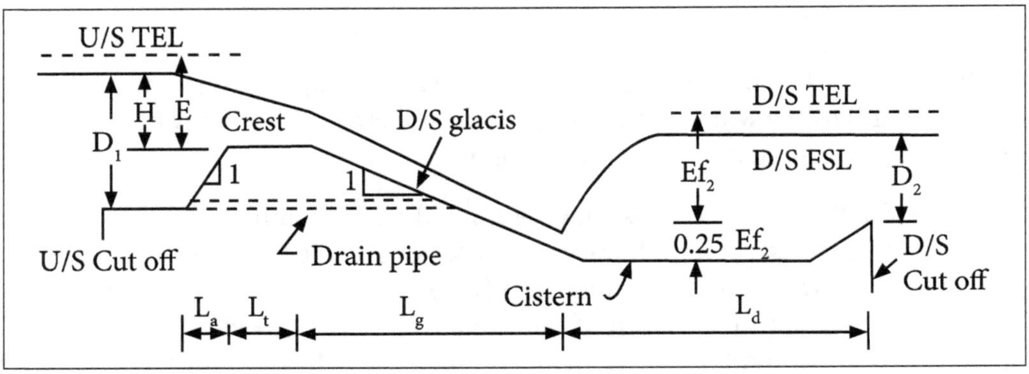

R.L. of cistern = d/s TEL -1.25 Ef2 = d/s FSL -1.25 D2

Length of cistern = 5 Ef2 for good earthen bed

or,

Ld = 6 Ef2 for erodible sandy soils.

The cistern is joined to the designed d/s bed with the up slope of 1 in 5 (1: 5) This arrangement enables the formation of hydraulic jump on the sloping glacis.

Provisions of Cut-offs

The cut-offs must be invariably provided at the upstream end of upstream glacis and also at the downstream end of the downstream cistern. The width of each curtain wall may be kept 0.4 m.

The depth can be as follows:

- The Depth of u/s cut-off = D1/3
- The Depth of d/s cut-off = D2/2
- The minimum depth must be 0.5 m.

Total Length of the Impervious Floor

The total length of the floor should be with the depth of curtain walls as fixed earlier give permissible exit gradient. Khosla's curve for the exit gradient can be used for this purpose.

The length of the floor between u/s and d/s cut-offs so determined .If it appears to be excessive then the downstream cut-off can be further deepened accordingly to achieve adequate floor length.

It may be noted that total impervious floor length comprises:

- Length of the cistern.
- Horizontal length of the d/s glacis.
- Crest length along the axis of the canal.
- Horizontal length of the u/s glacis.

In some cases little length still remains to be provided as per the earlier calculations .It may be provided on the u/s side of the u/s glacis.

Thickness of the Floor

The Minimum thickness on the u/s can be from 0.3 to 0.6 m. Floor thickness in the glacis and the cistern must be sufficient to withstand uplift pressure safely.

U/s Approach and U/s Protection

- When the fall combines with the functions of a discharge meter as well, then the side and bed approaches to the crest are necessarily to be gradual and smooth, so as to avoid eddies and impact losses. It also reduces concentration of flow.
- In non-meter falls, the side walls can be splayed at an angle of 45° from upstream edge of the crest. The walls are carried straight into canal berm for a length of at least 1 m.
- The bed approach can be by means of u/s glacis which has 1/2: 1 slope and joining tangentially the u/s end of crest having radius equal to E/2.

- Protection of bed and sides by stone or dry brick pitching can be done for a length of (D_1 + 0.5) m. The bed pitching can be laid at the slope of 1 in 10.

D/s Expansion and D/s Protection

- On downstream, parallel and vertical walls are provided upto the toe of the glacis.
- The expansion afterwards must be gradual so that expanding flow adheres to the sides and scour due to the formation of back-rollers on sides is also prevented. A rectangular hyperbolic expansion is given by Mitra's equation for hyperbolic expansion can be adopted.
- If this expansion works out too long, side splay of about 1 in 5 can be adopted. For small falls to effect economy expansion with side splay of 1 in 3 is considered to be sufficient.
- Side walls in expansion may be flared out from vertical to 1: 1, if the earth fill behind is not problematic like black cotton soil. In those cases the side walls can be designed as vertical gravity walls.
- Side protection having 20 cm thickness of dry brick pitching for a length of 3 D_2 must be provided. It should rest on the toe wall of 1½ brick thick and of depth equal to $D_2/2$ subject to minimum of 0.5 m depth.
- The deflector wall of height $D_2/10$ above d/s bed can be provided at the downstream end of cistern. The minimum height should be 15 cm. Thickness of the deflector wall can be kept 0.4 m.
- With provision of deflector wall at the end of the floor d/s bed pitching beyond the floor is not required.

Friction Blocks as Energy Dissipators

Friction blocks are known to be the most effective energy dissipators. In case of flumed straight glacis falls four rows of friction blocks can be provided. They are staggered in the plan. The u/s edge of the first row of the friction block is located at the distance of 5 times the height of the blocks (5.h) from the toe of glacis. The dimensions of the blocks may be as follows:

Let, height of the blocks = h

$h = D_1/8$

Length of the block = 3 h

Width of the block =2/3 h

Distance between rows = 2/3 h

When glacis is provided with baffle only two rows of friction blocks is essential upto 2 m fall. The u/s edge of the first row may be located at 1/3 length of d/s expansion from the end of the cristern floor.

3.2 Head and Cross Regulators and Design Principles

Head Regulator

The distributary head regulator was constructed at the upstream end of a channel where it takes off from main canal or a branch canal or a major distributary. The distributary head regulator must be distinguished from the canal head regulator which is provided at the canal headworks where a canal takes its supplies from the river source. The distributary head regulator serves to:

- Divert and regulate supplies into the distributary from the parent channel.
- Control the silt from entering the distributary from the parent channel.
- Measure the discharge that enters the distributary.

For this purpose of regulating the supplies entering the offtaking channel from the parent channel, abutments on either side of regulator crest are provided. Piers are placed along the regulator crest at the regular intervals. These piers and abutments have grooves for the purpose of placing planks or gates. The supplies into the offtaking channel are controlled by means of these planks or gates. The planks are used for small channels in which case the manual handling is possible. The span of hand-operated gates is also limited to 6 to 8 m. Mechanically-operated gates may, however, be as wide as 20 m.

An offtaking channel tends to draw excessive quantity of sediment due to the combined effects of the following:

- Because of their smaller velocities, lower layers of water are more easily diverted into the offtaking channels in comparison to the upper layers of water.
- Sediment concentration is generally much higher near the bed.
- Sediment concentration near the banks is usually higher because of the tendency of bottom water to move towards the banks due to difference in central and near -bank velocities of flow.

As such, when suitable steps are not taken to check the entry of excessive sediment into the off-taking channel, the offtaking channel will soon be silted up and might require repeated sediment removal.

Sediment entry into the offtaking channel may be controlled by causing the sediment to concentrate in the lower layers of water and then letting only the upper layers of water to enter into the offtaking channel.

Concentration of sediment in lower layers can be increased by providing smooth bed in parent channel upstream of the offtaking point. The smooth channel bed reduces turbulence which keeps sediment particles in suspension. In addition, steps which accelerate the flow velocity near the banks would also be useful. It must also be noted that the alignment of the off taking channel also affects the sediment withdrawal by the offtaking channel. Thus, the alignment of the offtaking distributary channel with respect to the parent channel needs careful consideration. The angle of offtake may be kept between 60° and 80° to prevent excessive sediment withdrawal by the offtaking channel. For the important works, the alignment of offtaking channels should be fixed on the basis of model studies.

For this purpose, it is essential to measure the discharge for which one can use gauge-discharge relationship of the distributary. Moreover, this relationship is likely to change with the change in the channel regime. Hence, it is a great advantage to use head regulator as a metering structure too.

Cross Regulator

A cross regulator is a structure constructed across a canal to regulate the water level in the canal upstream of itself and the discharge passing downstream of it for the following purposes:

- To feed offtaking canals located upstream of the cross regulator.
- To help water escape from canals in conjunction with escapes.
- To control water surface slopes in conjunction with falls for bringing canal to regime slope and section.
- To control discharge at an outfall of a canal into another lake or canal.

A cross regulator is generally provided downstream of an offtaking channel so that the water level upstream of the regulator may be raised, whenever necessary, to enable the offtaking channel draw its required supply even when the main channel is carrying low supply. The need of a cross regulator is necessary for all irrigation systems which supply water to distributaries and field channels by rotation and, therefore, require to provide full supplies to the distributaries even if the parent channel is carrying low supplies. Cross regulators can be combined with bridges and falls for economic and other special considerations.

Design Criteria for the Distributary Head Regulator and Cross Regulator

The effective waterway of a head regulator should not be less than 60 per cent of the width of the offtaking canal. It should be fixed such that the mean velocity of flow at full supply condition does not exceed 2.5 m/s. The overall waterway should at least be 70 per cent of the normal channel width of the offtaking channel at the downstream of head regulator. For cross regulators, waterway should be decided so that the resulting afflux does not exceed0.15 m.

The crest level of a head regulator should be such that the full supply discharge of the offtaking channel may be passed even when the parent channel is running with low supplies of the order of two-thirds of the fully supply discharge of the parent channel. It should be possible to maintain full supply level in the parent channel otect and retain the earth in the canal bank downstream of the offtake by means of a cross regulator. The water level at the location of offtake should be computed using back water computations. The level of the crest of the head regulator is obtained by subtracting the amount of head required from the computed water level at the offtake. In any case, the crest level should not be lower than the bed level of the offtaking channel. Usually, the crest level of the head regulator is 0.3 to 0.6 m higher than the crest level of the cross regulator. The amount of head over the crest of the head regulator, H is computed from the equation,

$$Q = CB_e H^{3/2}$$

The amount of head over the crest of the head regulator, H is computed from the equation,

Where, Q is the full supply discharge of the offtaking channel, and Be is the effective width of waterway and is given as:

$$B_e = B_t - 2\left(NK_p + K_\alpha\right)H$$

Here, Bt is the overall width of the waterway (i.e., the length of the crest), N the number of piers, and Kp and Ka are contraction coefficients for piers and abutments, respectively. Values of Kp range from 0.005 to 0.02 while those of Ka range from 0.1 to 0.2. C is a suitable discharge coefficient whose value can be taken as 1.84 for sharp -crested weirs (the crest width being less than 2/3 H) and 1.705 for broad-crested weirs (the crest width being greater than 2.5 H) for free flow conditions (3). If the flow is submerged, the values of C will have to be suitably modified depending upon the submergence ratio.

The crest of the cross regulator should be at least 0.15 m above the bed of the canal but should not be higher than 0.4 times the normal depth of the upstream canal (4).

The crest width should be greater than 2/3 H and should be sufficient to accommodate the gate cill. The upstream and downstream slopes of the crest can be at slopes of 2 (H): 1 (V). Impervious floor and cutoffs will be designed from considerations of hydraulic jump, uplift pressures, safe exit gradient, and scour depth. Similarly 'flexible' protection works at the upstream and downstream ends of the impervious floor will be provided in the form of block protection, inverted filters and launching apron.

3.3 Cross Drainage Works: Types and Selection

It is a structure for carrying the discharge from a natural stream across the canal intercepting the stream. Canal comes across obstructions like rivers, natural drains and other canals.

Several various types of structures that are built to carry canal water across the above mentioned obstructions or vice versa are known as cross drainage works.

There are many different factors that are involved in selecting a specific type of Cross drainage works and in selecting a suitable site for cross drainage works.

It is generally a costly item and must be avoided by diverting one stream into another stream.

Changing the alignment of canal so that it crosses below the junction of two streams.

Types of Cross Drainage Works

Depending upon the levels and the discharge, it can be of the following types: Cross drainage works carrying a canal across the drainage.

Structures that fall under this type are as follows:

- An Aqueduct.
- Siphon Aqueduct.

Aqueduct

When HFL of the drain is sufficiently below the bottom of canal in a manner that the drainage water flows freely under gravity, the structure is known as Aqueduct.

In this, canal water is being carried across the drainage in a trough supported on the piers.

Bridge carrying water provided when necessary level difference is available between the canal and canal bed is sufficiently higher than HFL.

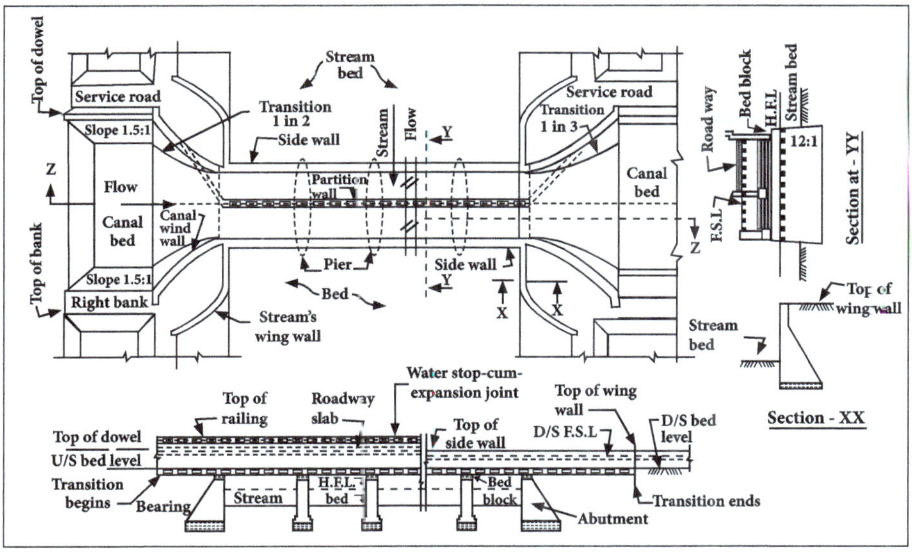

Typical section of an aqueduct.

Crossing works (Aqueduct).

Siphon Aqueduct

In case of the siphon Aqueduct, the HFL of drain is higher above the canal bed, and water runs under the siphonic action through the aqueduct barrels.

Siphon Aqueduct.

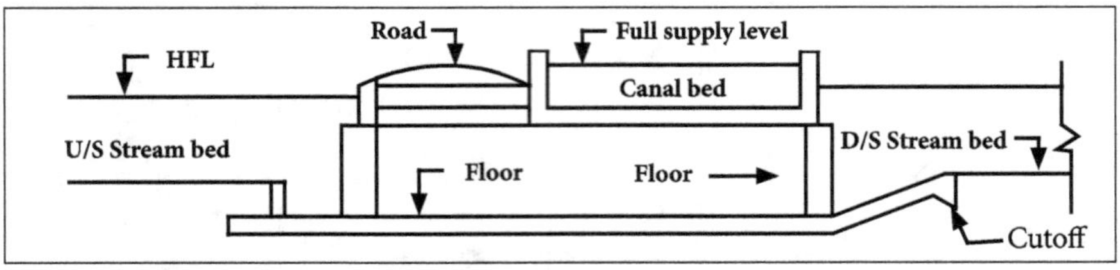

Layout of siphon aqueduct.

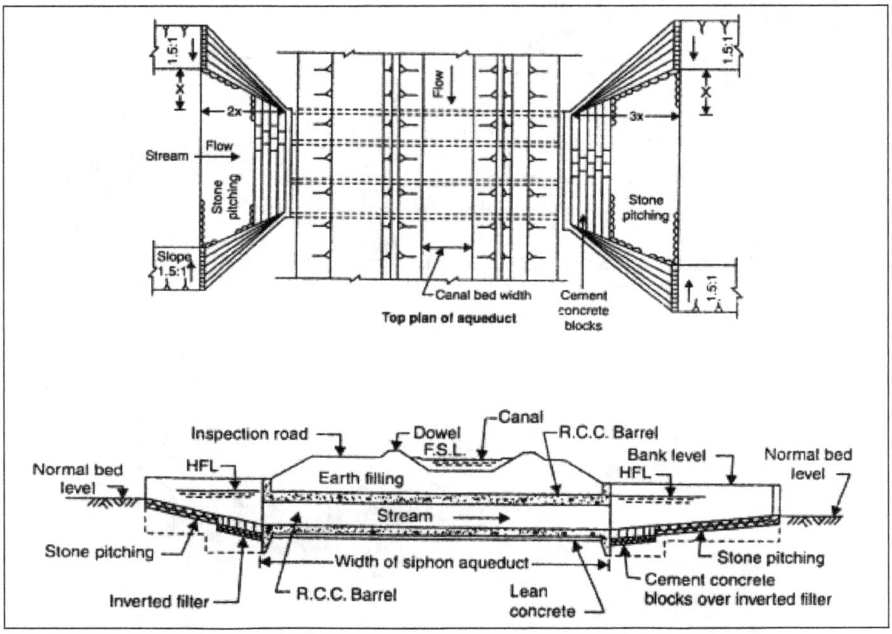

Typical section of siphon aqueduct (type I).

The drain bed is generally depressed and provided with pucci floors on the upstream side. Drainage bed may be joined to the pucca floor either by a vertical drop or by glacis of 3:1. The downstream rising slope must not be steeper than 5:1.

When canal is passed over the drain, the canal remains open for inspection throughout and damage caused by flood is rare.

During heavy floods, the foundations are susceptible to scour or waterway of drain may get choked due to debris, tress etc.

Type I: In this type of structure, earthen canal banks are carried as such and, therefore, the culvert length has to be long enough to support the water section as well as the earthen banks of the canal. In this type of structure, canal section is not flumed and thus remains unaltered. Therefore, the width of the structure is maximum. This type of structure, obviously, saves on the canal wings and bank connections and is justified only for small streams so that the length of the structure is small. An extreme example of this kind of structure would be when the stream is carried by means of a pipe laid under the bed of the canal.

Type II: This type of structure is similar to the Type I with a provision of retaining the walls to retain the outer slopes of earthen canal banks. This reduces the length of culvert. This type of construction may be considered suitable for the streams of intermediate sizes.

Type III: In this type of structure, the earthen canal banks are discontinued through aqueduct and the canal water is carried in a trough which may be of either masonry or concrete. The earthen canal banks are connected to the corresponding through walls on their sides by means of wing walls. The width of canal is also reduced over crossing. In this type of structure, width of the structure is minimum and, thus, the structure is suitable for large streams requiring considerable length of aqueduct between the abutments.

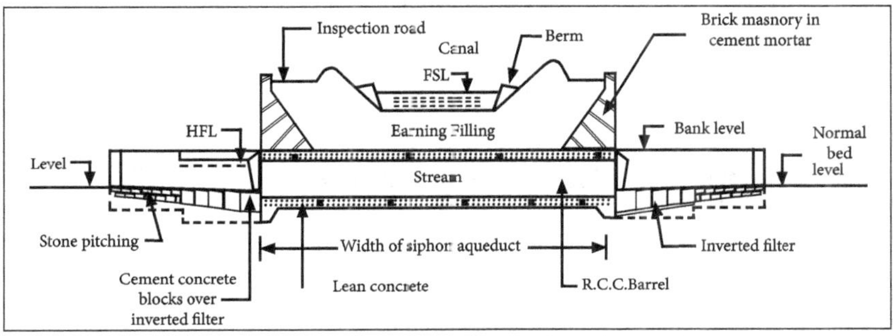

Typical section of siphon aqueduct (type II).

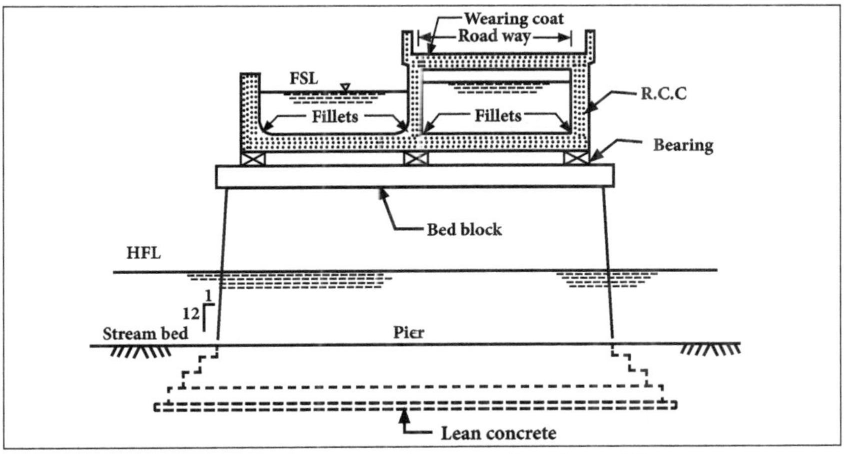

Typical section of siphon aqueduct (type III).

Selection of Suitable Site for Cross Drainage Works

The factors that affect the selection of suitable type of cross drainage works are:

- Relative bed levels and water levels of canal and the drainage.
- Size of the canal and drainage.

- The following considerations are important.
- When the bed level of the canal is well above the HFL of the drainage, an aqueduct is the obvious choice.
- When the bed level of the drain is well above FSL of canal, super passage is provided.
- The necessary headway between canal bed level and the drainage HFL can be increased by shifting the crossing to the downstream of drainage. If, however, it is not possible to change canal alignment, a siphon aqueduct can be provided.
- When canal bed level is much lower, but the FSL of canal is present higher than the bed level of drainage, a canal siphon is preferred.
- When the drainage and canal cross each other practically at the same level, a level crossing can be preferred. This type of work is avoided as far as possible.

Factors which Influence the Choice or Selection of Cross Drainage Works

The considerations which govern the choice between the aqueduct and the siphon aqueduct are:

- Suitable canal alignment.
- Suitable soil available for bank connections.
- Nature of available foundations.
- Permissible head loss in canal.
- Availability of funds.

Compared to an aqueduct, a super passage is inferior and must be avoided whenever possible. Siphon aqueduct is preferred over siphon unless large drop in drainage bed is needed.

3.3.1 Design Principles of Aqueduct, Siphon Aqueduct and Super Passage

1. Estimation of Design Flood Discharge of a Drain: The drain to be crossed can be small or like a river. In all cases correct assessment of maximum flood or peak flow of the drain must be obtained beforehand.

Waterway Requirement for a Drain: Lacey's regime perimeter equation holds good basis for calculating drainage waterway. The equation is given by,

$$Pw = 4.825 Q1/2$$

Where, Pw is waterway to be provided for drain at the site in metres. Q is flood discharge of drain in m3/sec. As piers reduce the actual waterway available, the length between abutments (Pw) can be increased by 20 per cent. When the waterway is fixed from the Lacey's regime perimeter equation, regime condition in the drain upstream and the downstream of the structure is not disturbed appreciably. To confine drainage water to the desired waterway guide banks can be constructed.

Velocity of Flow through Barrel: The velocity of flow through the barrel may range from 1.8 m/sec to 3 m/sec, the reason for selection of this range is that the lower velocities can cause silting in the barrels, Whereas when the velocity is higher than 3 m/sec the bed load can cause abrasion of the barrel floor and subsequently it can be damaged.

Height of Opening: Once the waterway discharge and velocity are fixed the depth of flow will be obtained easily. There must be sufficient headway or clearance left between the HFL and the bottom of canal bed. A clearance of 1 m or half the height of the culvert, whichever is less will be sufficient. Thus, Height of opening = Depth of flow + Clearance or headway.

Number of Spans: After determining the total length of an aqueduct between the abutments number of spans to be provided may be fixed on the basis of the following two considerations:

- Structural strength required.
- Economical consideration.

For example, when the arches are used the number of spans to be given can be more. When the cost of construction in the foundation is high, small number of spans is adopted and then RCC beams can be used.

Canal Waterway: Generally fluming ratio is taken to be 1/2. This ratio is adopted in such a way that the velocity of flow in trough does not go above critical velocity limit. Generally velocity of flow must not be more than 3 m/sec. This precaution is taken to avoid the possibility of formation of hydraulic jump. The obvious reason is that when hydraulic jump is formed it will absorb energy. In this process valuable head is lost and large stresses are produced in the structure.

Length of the Contraction or Approach Transition: Once the width at throat is fixed, length of contraction may be determined after knowing convergence ratio. The convergence ratio is generally taken as 2: 1, i.e., not steeper than 30°.

Length of the Expansion or Departure Transition: Length of the expansion on the downstream side of the aqueduct can be fixed after knowing the expansion ratio. The expansion ratio is generally taken as 3: 1, i.e., not steeper than 22.5°. To maintain the streamlined flow and also to reduce loss of head the transitions are generally made up of the curved and flared wing walls.

The design of the transition can be worked out by making use of any of the following three methods:

- Hind's method.
- Mitra's hyperbolic transition method.
- Chaturvedi's semi-cubical parabolic transition method.

It can be noted that whereas Hind's method may be used when the water depth in the normal section and the flumed trough also varies, the remaining two methods may be used only when the water depth remains to be constant in the normal canal section as well as the trough section.

Bank Connections: An aqueduct requires four sets of wing wall, Canal wing walls on the upstream and downstream side of aqueduct protect and retain the earth in canal banks. The foundation of canal wing walls must not be left in the embanked earth. The wing walls must be based on the sound foundation in the natural ground. In the transitions the side slopes of the natural section are warped to conform with shape of the trough over the drain.

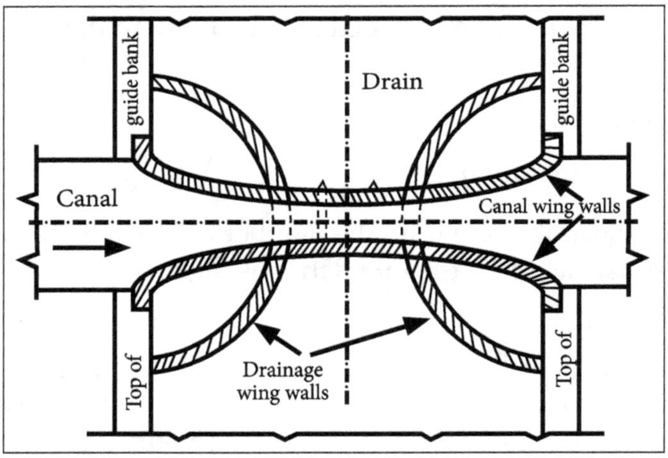

Drainage wing walls are provided on the upstream and downstream of the barrel to protect and to retain the natural sides of the drain. As the bed of the drain gets scoured during floods the drainage wing walls need to be taken deep into the foundation below maximum scour depth. The wing walls should be taken back essentially into the top of guide banks. The wing walls must be designed to permit the smooth entry and exit of the flow in the drain.

Hind's Method for Design of Transition

This method is based on the premise that there is a minimum loss of head, flow is streamlined and normal flow conditions in the canal are restored before the canal discharges pass down on to the earthen section immediately after curved and in flared transitions.

In the contraction or approach transition, the throat portion and the expansion or departure transition are shown. It can be seen that sections 1-1, 2-2, 3-3 and 4-4 indicate the start of contraction, end of contraction, start of expansion and end of expansion respectively.

Thus the contraction or approach transition lies between sections 1 and 2, throat between the sections 2 and 3 and expansion or departure transition between sections 3 and 4. Upto section 1 and beyond section 4 canal flows under its normal conditions and therefore the canal parameters at these two points are equal and already known. So also conditions of flow and canal parameters are same between sections 2 to 3 which represents throat or trough portion.

The design procedure can be outlined as follows:

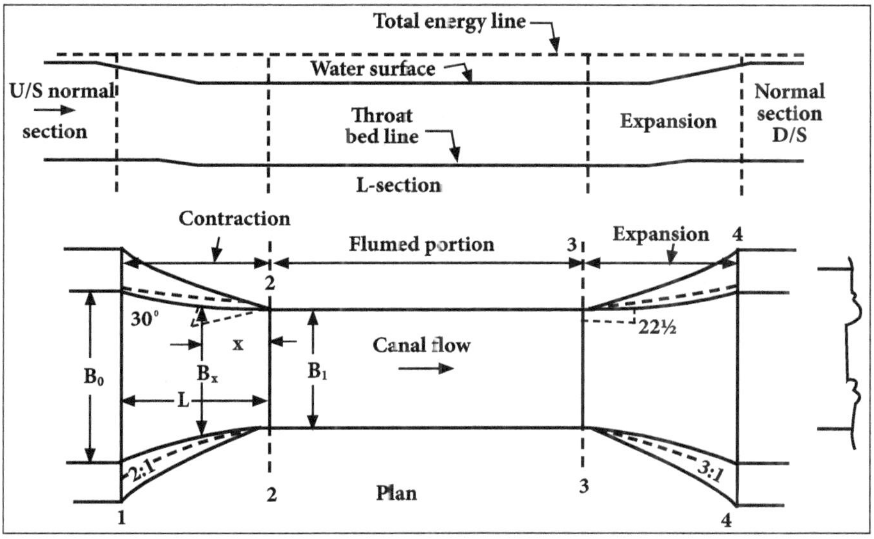

Let D and F with appropriate subscripts refers to depths and velocities at four sections also since canal levels and dimensions are already known in the section 4-4:

Step 1: TEL at section 4-4 =Elevation of Water surface + V24/2g where water surface elevation at sec. 4-4 = Bed level + D4.

Step 2: TEL at sec. 3-3 = (TEL at sec 4-4) + (energy loss between sec. 3 and 4) Energy loss between the sections 3-3 and 4-4 takes place due to the expansion of streamlines and also due to friction. Neglecting loss due to the friction which is small and taking loss due to the expansion to be,

$$0.3\left(\frac{V_3^2 - V_4^2}{2g}\right)$$

TEL at sec. 3-3 = (TEL at sec. 4-4) + $0.3\left(\frac{V_3^2 - V_4^2}{2g}\right)$

Therefore, the Water surface level at sec. 3.3 = (TEL at sec. 34) - $\left(\frac{V_3^2}{2g}\right)$

And the bed level at sec 3-3 = (Water surface level at sec. 3-3) – (D_3).

Step 3: Similarly, TEL at section 2-2 = (TEL at sec. 3-3) + (Head loss between the sections 2 and 3) between sections 2-2 and 3.3 canal flows through trough section of same dimension and therefore flow is now uniform. Loss of head in the trough is only due to the friction losses. The loss of head can be calculated using Manning's equation.

$$Q = A \cdot \frac{1}{N} R^{2/3} \cdot S^{1/2}$$

Water surface level at sec. 2-2 = (TEL at sec. 2-2) - $\left(\frac{V_2^2}{2g}\right)$.

Bed level at sec 2-2 = (Water surface level at sec. 2.2) - (D_2).

Since depth and velocity are same in the trough from sections 2 to 3 it may be noted that the total energy line, water surface line and bed line are parallel to each other.

Step 4: On the similar lines.

TEL at sec. 1.1 = (TEL at sec. 2-2) + (Head loss between secs. 1 and 2).

Between sections 2.2 and 1.1 energy loss takes place due to contraction of streamlines and friction. Neglecting friction loss, loss due to contraction may be taken to be $0.2\left(\frac{V_2^2 - V_1^2}{2g}\right)$,

TEL at sec. 1-1 = (TEL at sec. 2.2) + $0.2\left(\frac{V_2^2 - V_1^2}{2g}\right)$.

Water surface level at sec. 1-1 = (TEL at sec. 1-1) - $\left(\frac{V_1^2}{2g}\right)$ and 2 bed level at sec. 1-1 = (Water surface level at sec. 1 1)-(D1).

Step 5: As mentioned in the first four steps the bed level, water surface level and level of total energy line can be determined at the four sections.

Now the TE line, water surface line and the bed line can be drawn as follows:

- Now total energy line can be drawn by joining these points at four sections by a straight line.
- The bed line can also be drawn as straight lines between adjacent sections if the fall or rise of bed level is small. The corners must be rounded off. In case the drop in the bed line is appreciable the bed lines must be joined with a smooth tangential reverse curve.

- It is now clear that between any two consecutive sections the drop in water surface level may result due to (i) drop in the TE line between the two sections; (ii) increased velocity head in the contraction; and (iii) decreased velocity head in the expansion.

This drop in water surface gets negotiated by two parabolic curves. As shown in Figures below for contraction and expansion, this is achieved by convex upwards curve followed by concave upwards curve in former transition and concave upwards curve followed by the convex upwards curve in the latter transition.

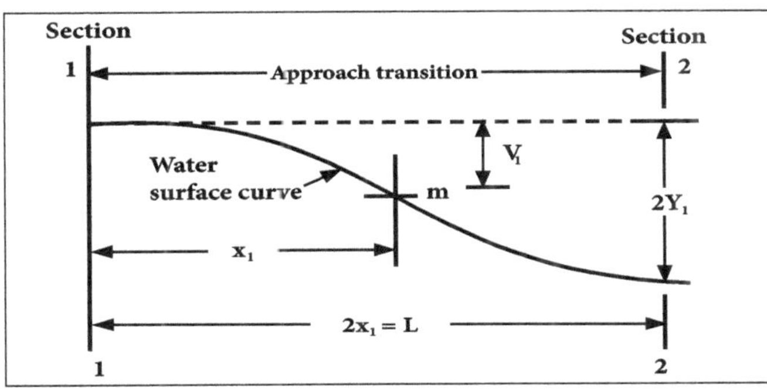

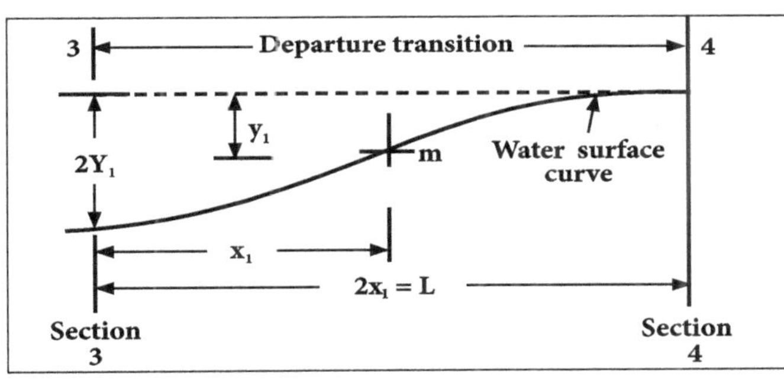

It may be seen from the above figures.

L = Length of transition= $2x_1$ and, $2y_1$ = Total drop or rise in the water surface. The point m is the midpoint of transition length and is situated to divide the total drop as well as length equally.

Taking water surface at the section point as the origin equation of the parabola is given by,

$$y = cx_2$$

Substituting the known values of y_1 and x_1,

$$c = y_1/x_2$$

With this value of c parabolic water surface curves can be plotted starting from section points which represent the origin.

The equation is to be used for plotting now reduced to,

$$y = (y_1/x_{12}) . x_2$$

Thus, water surface profile can be plotted.

Step 6: Velocity and area of flow at various points can be determined.

(i) Velocity head at any point is given by the difference between the T.E.L. and the water surface.

Velocity head,

$$hv = TEL - WS \text{ line}$$

Also,

$$= hv = v_2/2g$$

So velocity (V) at each point = $\sqrt{2g.hv}$

(ii) The area of flow at any point may now be obtained by using the simple formula,

$$A = Q/V$$

With known values of A and D other dimensions of the trapezoidal channel can be calculated using the following formula,

$$A = BD + SD2$$

Where B is bed width and S: 1, i.e., (H: V) is the side slope.

In case of flared wing walls, the side slopes are gradually brought to vertical from an initial slope. The value of side slope at any intermediate section in the transition length may be interpolated in proportion to the length of the transition achieved upto that point.

Mitra's Hyperbolic Transition Method

This method is based on the principle that:

- Along with the discharge, the depth of flow in the canal is also a constant.
- Rate of change of velocity per unit length of the transition is also constant throughout the length of transition.

From above figure it can be seen that,

Bo = normal bed width of canal;

Bt = bed width in the throat or trough;

Bx = width at any distance x from the extremity of the trough; and L = total length of the transition.

Also let V_o V_t and V_x be the velocities at the corresponding sections.

From second principle,

$$\frac{V_t - V_x}{x} = \frac{V_t - V_o}{L} \qquad ...(1)$$

From first principle,

$$Q = V_t \times B_t \times D$$
$$= V_x \times B_x \times D = V_o \times B_o \times D$$

Since Q and D are constant,

Let,

$$\frac{Q}{D} = K$$
$$\therefore K = V_t \times B_t = V_x \times B_x = V_o \times B_o$$

or,

$$V_t = \frac{K}{B_t}$$
$$V_x = \frac{K}{B_x}$$

And,

$$V_o = \frac{K}{B_o}$$

Substituting these values in equation (1) above,

$$\frac{\left(\frac{K}{B_t} - \frac{K}{B_x}\right)}{x} = \frac{\left(\frac{K}{B_t} - \frac{K}{B_x}\right)}{L}$$

or,

$$B_x = \frac{B_o B_t B}{LB_o - (B_o B_t)x}$$

Thus B_x gives bed width at any distance x from the throat.

Chaturvedi's Semi-cubical Parabolic Transition Method,

It states that $x = \frac{L B_o^{3/2}}{B_o^{3/2} - B_t^{3/2}} \left\{ 1 - \left(\frac{B_t}{B_x} \right)^{3/2} \right\}$

Design Principles for Siphon Aqueduct

It is clear that the siphon aqueducts are basically different from ordinary aqueducts. As such criteria for aqueduct design is not sufficient in the design of siphon aqueducts.

In addition to the above considerations the following criteria should be adopted while designing the siphon aqueducts:

1. Discharge through Siphon Barrel

The head which causes the flow (it also represents head loss in barrel) through the inverted siphon barrel may be obtained from Unwin's formula,

$$h = \left(1 + f_1 + f_2 \frac{L}{R} \right) \frac{V^2}{2g} - \frac{V_\alpha^2}{2g}$$

Where h is the head causing flow, it is the loss of head in the barrel in m.

L is the length of barrel in m.

R is hydraulic mean radius of the barrel in m.

V is velocity of flow through the barrel in m/sec.

Va is velocity of the approach in m/sec, it is generally neglected.

f_1 is a coefficient for loss of head at the entry and generally taken as 0.505.

f_2 is a coefficient which accounts for the friction in the barrel.

$$f_2 = \alpha \left(1 + .305 \frac{b}{R} \right)$$

Where a and b are constants.

When the velocity of approach is small, the term representing the loss of head in friction in the barrel $\left(\text{i.e.,} f_2 . \frac{L}{R} . \frac{V^2}{2g} \right)$ may be neglected.

Velocity of flow through the barrel in generally limited to 2 to 3 m/sec.

Thus, All the values are known head loss in barrel or head causing flow may be calculated. This value when added to the High Flood Level (HFL) on d/s of the aqueduct gives u/s HFL.

Adding these free board to u/s HFL we can get the top of river protection works like guide bunds and marginal bunds.

2. Uplift Pressure on the Roof of the Barrel

As the barrel runs full during floods, there exists positive pressure in the barrel. Due to the positive pressure in the barrel, the roof is subjected to uplift pressure. The uplift pressure diagram for the roof can be drawn knowing the pressure head on the u/s and d/s side of the barrel.

Pressure head on the d/s side of the barrel is equal to the height of the water level above the bottom of the roof. Pressure head on the u/s side may be obtained by adding the loss of head in barrel to the pressure head on d/s side. The loss of head may be obtained from the Unwin's formula. Figure below shows the profile of hydraulic gradient line that now exist. It can be seen that maximum uplift pressure occurs at the u/s end of the barrel roof.

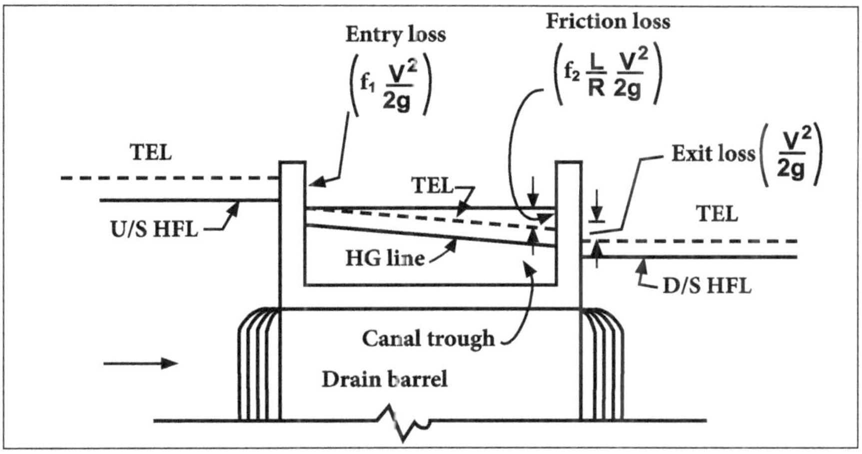

While designing the trough it is necessary to consider two extreme conditions, namely:

- The barrel runs full during maximum flood and there is no water in the canal trough. This condition gives maximum uplift pressure acting on the trough.
- The canal trough is carrying the full discharge but the barrel does not run full and thus there is no uplift on the roof of the barrel.

In order to limit the thickness of the trough it is advisable to provide reinforced concrete roof with reinforcement at the bottom to take the load of the canal trough and the reinforcement at top to resist uplift pressure by bending.

3. Uplift Pressure on the Floor of the Barrel

Unlike other hydraulic structures aqueducts are subjected to the two different types of uplift pressures from the two different sources. They are the following:

Static Uplift Pressure Due to Rise in Water Table: The water table rises many times up to bed level of drain. Specifically in case of siphon aqueduct whose floor bed is depressed below bed of the drainage, the static uplift pressure acts on floor bed. The uplift pressure is similar to the difference of the bed level of the drain and that of floor level of the barrel.

Uplift Pressure Due to Seepage of Canal Water to the Drain: Since there exists the difference of level between canal water level and the drainage water level seepage flow takes place where conditions are desirable. The seepage head is maximum when the canal runs with full capacity and there is no flow in the drain below. As shown in given Figure below, seepage flow is not simple but flow pattern is found three dimensional everywhere. The seepage flow starts from either sides of impervious canal trough bed and reappears on either sides of the impervious barrel floor in the drain.

Since no approximation to two dimensional flow is possible, Khosla's theory cannot be applied strictly. Solution by intricate "method of relaxations" is still possible but it is too elaborate. For design purposes Bligh's creep theory principle explained below can be applied. For major works, however, it is essential to check the results of the preliminary design so obtained by the model studies.

Taking the case of the first barrel where seepage will be maximum, total creep length – (creep length a-b) + (creep length b-c),

$$L = L_1 + L_2$$

Total seepage head = canal FSL – d/s bed level of drain = Hs

Residual seepage head at b = -Hs/L x L_2

Total residual seepage head at b can be considered to design the thickness of the entire floor of all the barrels.

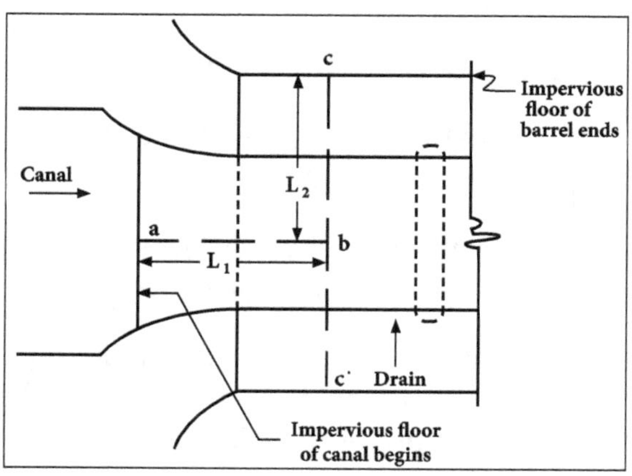

The floor thickness of the barrel is in fact designed considering total uplift pressure created by static uplift condition and canal seepage flow mentioned above.

In order to limit the thickness of the floor RCC construction can be adopted as then part of the pressure is resisted by weight of floor and remaining by bending strength of the floor. In such arrangement the pressure gets transferred to the piers and is resisted by the entire weight of the superstructure.

When it is seen that uplift pressure is very high it can be reduced by providing suitable safe guards.

They are:

- Increasing the length of the impervious floor of the canal bed so that creep length is also increased;
- Providing drainage holes or relief holes in the floor of the barrel in conjunction with inverted filter below the floor. To avoid choking of the relief holes and filter below the drain silt relief holes must be provided with flap valves.

Super Passage

The hydraulic structure in which the drainage is passing over the irrigation canal is termed as super passage. This structure is suitable when the bed level of drainage is above the flood surface level of the canal. The water of the canal passes clearly below the drainage:

- A super passage is same to an aqueduct, except in this case the drain is over the canal.
- The FSL of the canal is lower than the underside of the trough carrying drainage water. Thus, the canal water runs under the gravity.
- Reverse of an aqueduct.

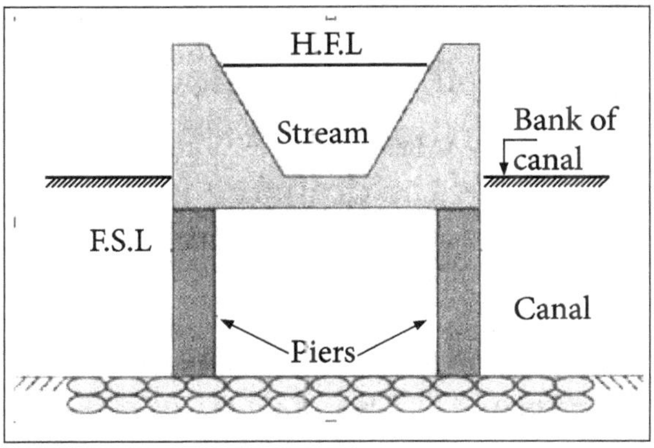

Super passage.

3.4 Outlets: Types, Proportionality, Sensitivity and Flexibility

Canal Outlet

Canal outlets also called farm turnouts in some countries are structures at the head of a water course or field channel.

Supply canal is usually under the control of an irrigation authority under the State government. Since an outlet is a link connecting the government owned supply channel and the cultivator owned field channel, the requirements should satisfy the needs of both the groups.

Equitable distribution of the canal supplies is dependent on the outlets it must not only pass a known and constant quantity of water but must also be able to measure the released water satisfactorily.

Since the outlets release water to each and every farm watercourse such structures are more numerous than any other irrigation structure.

It is essential to design an outlet in such a way that it is reliable and be also robust enough such that it is not easily tampered with.

Cost of an outlet structure should be low and should work efficiently with a small working head. Since a larger working head would require higher water level in the parent channel resulting in high cost of the distribution system.

Discharge through an outlet is usually less than 0.085 cumecs. Various types of canal outlets have been evolved from time to time but none has been accepted as universally suitable.

It is very difficult to achieve a perfect design fulfilling both the properties of 'flexibility' as well as 'sensitivity' because of various indeterminate conditions both in the supply channel and the watercourse of the following factors:

- Discharge and silt.
- Capacity factor.
- Rotation of channels.
- Regime condition of distribution channels, etc.

When the canal water has reached near the fields to be irrigated, it has to be transferred to the watercourses. At the junction of the watercourse and the distributary, an outlet is provided. An outlet is a masonry structure through which water is admitted from the distributary into a watercourse. It also acts as a discharge measuring device. The discharge though an outlet is usually less than 0.085 m^3/s (3). It plays a vital role in the warabandi system of distributing water. Thus, an outlet is like a head regulator for the watercourse. The main objective of providing an outlet is to provide ample supply of water to the fields, whenever needed. If the total available supply is insufficient, the outlets must be such that equitable distribution can be ensured. The efficiency of an irrigation system depends on the proper functioning of canal outlets which should satisfy the following requirements:

- The outlets must be strong and simple with no moving parts which would require periodic attention and maintenance.
- The outlets should be tamper-proof and if there is any interference in the functioning of the outlet, it should be easily detectable.
- The cost of outlets should be less as a large number of these have to be installed in an irrigation network.
- The outlet should be able to draw sediment in proportion to the amount of water withdrawn so that there is no silting or scouring problem in the distributary downstream of the outlet.
- The outlets should be able to function efficiently even at low heads.

The choice of type of an outlet and its design are governed by factors such as water distribution policy, water distribution method, method of water assessment, sources of supply, and the working of the distributary channel. Water may be distributed on the basis of either the actual area irrigated in the previous year or the actual culturable command area. The discharge from the outlet should be capable of being varied in the first case, but, can remain fixed in the second. The method of water distribution may be such that each cultivator successively receives water for a duration in proportion to his area. Alternatively, all the cultivators share the outlet discharge simultaneously. The first system is better as it results in less loss of water. The outlet capacity is decided keeping in view the method of water distribution. If the assessment is by volume, the outlet discharge should remain constant and not change with variation in the water levels of the distributary and the watercourse. On the other hand, if water charges are decided on the basis of area, the variation in the outlet capacity with water levels of the distributary and watercourse is relatively immaterial. With a reservoir as supply source, the cultivators can be provided water whenever needed and, hence, the outlets should be capable of being opened or closed. The outlets generally remain open if the supply source is a canal without storage so that water is diverted to the field when the canal is running. At times, the amount of water in the main canal may not be sufficient to feed all the channels simultaneously to their full capacity. As such, either all the channels may run with low discharge or groups of channels may be supplied their full capacity by rotation. In the first case, the outlets must be able to take their proportionate share even with large variations in the discharge of the distributary channel. In the second case, the outlets must be such that the required amount of water is available for all the channels being fed with their full capacity. It should be noted that whereas the cultivator prefers to have outlets capable of supplying constant discharge, the canal management would prefer that the outlets supply variable discharge depending upon the discharge in the distributary channel so that the tail end of the channel is neither flooded nor dried. Obviously, both these requirements cannot be fulfilled simultaneously.

Types of Outlet Canal outlets are of the following three types:

- Non-modular outlets.
- Semi-modular outlets.
- Modular outlets.

Non-modular Outlets

These modules are classified in three types, which are as follows:

These outlets operate in such a way that the flow passing through them is a function of the difference in water levels of the distributing channel and the watercourse. A variation in either affects the discharge. These outlets consist of regulator or circular openings and pavement.

The effect of downstream water level is more with short pavement.

They are those whose discharge capacity depends on the difference of water levels in the distributary and the watercourse. The discharge through non-modular outlets fluctuates over a wide range with variations in the water levels of either the distributary or the watercourse. Such an outlet is controlled by a shutter at its upstream end. The loss of head in a non-modular outlet is less than that in a modular outlet. Hence, non-modular outlets are very suitable for low head conditions. However, in these outlets, the discharge may vary even when the water level in the distributary remains constant. Hence, it is very difficult to ensure equitable distribution of water at all outlets at times of keen demand of water.

Semi-modular Outlets

The discharge through a semi-modular outlet (or semi-module or flexible outlet) depends only on the water level in the distributary and is unaffected by the water level in the watercourse provided that a minimum working head required for its working is available. A semi-module is more suitable for achieving equitable distribution of water at all outlets of a distributary. The only disadvantage of a semi-modular outlet is that it involves comparatively greater loss of head.

Modular Outlets

They are those whose discharge is independent of the water levels in the distributary and watercourse, within reasonable working limits. These outlets may or may not have moving parts. In the latter case, these are called rigid modules.

Though modular outlets like the Gibb's module have been designed and implemented earlier they are not very common in the present Indian irrigation engineering scenario.

Modular outlets with moving parts are not simple to design and construct and are, hence, expensive. A modular outlet supplies fixed discharge and, therefore, enables the farmer to plan his irrigation accordingly. However, in case of excess or deficient supplies in the distributary, the tail-end reach of the distributary may either get flooded or be deprived of water. This is due to the reason that the modular outlet would not adjust its discharge corresponding to the water level in the distributary. But, if an outlet is to be provided in a branch canal which is likely to run with large fluctuations in discharge, a modular outlet would be an ideal choice. The outlet would be set at a level low enough to permit it to draw its due share when the branch is running with low supplies. When the branch has to carry excess supplies to meet the demands of the distributaries, the discharge through the modular outlet would not be affected and the excess supplies would reach up to the desired distributaries. Similarly, if an outlet is desired to be located upstream of a regulator or a raised crest fall, a modular outlet would be a suitable choice.

Parameters for Studying the Behaviour of Outlets

Flexibility

F: Flexibility is defined as the ratio of the rate of change of discharge of the outlet to the rate of change of discharge of the distributary channel.

$$F = \frac{dq/q}{dQ/Q}$$

Where,

F = Flexibility of the outlet

q = Discharge passing through the outlet

Q = Discharge in the distributary channel

If,

H = the head acting on the outlet,

$$q = CH^m$$

Where, C and m are constants depending upon the type of outlet

If,

y = the depth of water in the distributary,

$$Q = Ky^n$$

Where, k and n are constants,

$$\frac{dq}{dH} = CmH^{m-1} = \left(CH^m\right)\times\left(m/H\right) = q\times\frac{m}{H}$$

$$\Rightarrow \frac{dq}{q} = \frac{m}{H}\times dH$$

Again,

$$\frac{dQ}{Q} = Kny^{n-1} = \left(Ky^n\right)\times\left(n/y\right) = Q\times\frac{n}{y}$$

$$\Rightarrow \frac{dQ}{Q} = \frac{n}{y}\times dy$$

Thus,

$$F = \frac{\frac{m}{H}\times dH}{\frac{n}{y}\times dy} = \frac{m}{n}\times\frac{y}{H}\times\frac{dH}{dy}$$

A change in water depth of the distributary (dy) would result in an equal change in the head working on the outlet (dH) so that,

$$dy = dh$$

So,

$$F = \frac{m}{n}\times\frac{y}{H}$$

Proportionality

The outlet is said to be proportional when the rate of change of outlet discharge equals the rate of change of channel discharge.

Thus,

$$\frac{dq}{q} = \frac{dQ}{Q}$$

So,

$$F = 1, \text{ i.e. } \frac{m}{n} \times \frac{y}{H} = 1$$

$$\Rightarrow \frac{H}{y} = \frac{m}{n} = \frac{\text{Outline index}}{\text{Channel index}}$$

The outlet is said to be sub-proportional, if F < 1.

or,

$$\frac{H}{y} > \frac{m}{n}$$

The outlet is said to be hyper-proportional, if F > 1.

or,

$$\frac{H}{y} < \frac{m}{n}$$

Setting

It is the ratio of the depth of the silt level of the outlet below the FSL of the distributary, to the full supply depth of the distributary,

$$\text{setting} = \frac{H}{y}$$

For proportional outlet, setting $= \frac{H}{y} = \frac{m}{n}$

For a wide trapezoidal channel, the discharge is proportional to y5/3, so, n = 5/3

Discharge through an orifice type outlet is proportional to H1/2, so, m =1/2

Thus,

$$\text{Setting} = \frac{H}{y} = \frac{m}{n} = \frac{1/2}{5/3} = \frac{3}{10} = 0.3$$

For a weir type outlet, the discharge is proportional to H3/2 Hence, the setting for a combination of a weir type outlet and a trapezoidal channel,

$$= \frac{m}{n} = \frac{3/2}{5/3} = \frac{9}{10} = 0.9$$

Thus an orifice or a weir type outlet shall be proportional, if the outlet is set at 0.3 and 0.9 times depth below the water surface respectively.

Sensitivity, S: It is defined as the ratio of the rate of change of discharge through the outlet to the ratio of change of water level of the distributary.

$$S = \frac{\frac{dq}{q}}{\frac{dG}{y}}$$

Relation between Sensitivity and Flexibility,

$$F = \frac{\frac{dq}{q}}{\frac{dQ}{Q}}$$

But,

$$\frac{dq}{q} = \frac{n}{y} \times dy$$

$$\therefore F = \frac{\frac{dq}{q}}{\frac{n}{y} \times dy} = \frac{1}{n} \times \frac{\frac{dq}{q}}{\frac{dy}{y}}$$

Since,

$$dg = dy$$

So,

$$F = \frac{1}{n} \times S$$

Thus,

$$S = n \times F$$

For rigid modules, the discharge is fixed, and hence sensitivity is zero. The greater the variation of discharge through an outlet for a given rise or fall in water level of the distributary, the larger is the sensitivity of the outlet.

3.5 River Training: Objectives and Approaches

River training structures for canal headworks are required for the following purposes:

- To prevent outflanking of the structure.

- To minimise possible cross-flow through the barrage or weir which may endanger the structure and protection works.
- To prevent flooding of the riverine lands upstream of the barrages and weirs.
- To provide favourable curvature of flow at the head regulator from the consideration of entry of sediment into the canal.

The following types of river training structures are usually provided for weirs:

- Guide banks.
- Approach embankments.
- Afflux embankments.
- Groans or spurs.

To ensure smooth and axial flow of water, to prevent the river from out the works due to change in its course.

River Training Works.

Guide Banks: Earthen embankments => stone pitching

Force the river into restricted channel, to ensure almost axial flow near the weir site. (Embankments in continuation of G-Banks. To contain flood within flood plains).

Marginal Bunds: Provided on the upstream in order to protect the area from submergence due to rise in HFL, caused by afflux.

Groans or Spurs: Embankment type structures constructed transverse to river flood, extending from the banks into the river (also transverse dykes).

Protect the bank from which they are extended by deflecting the current away from the bank.

The purpose of guide banks is to narrow down and restrict the course of a river so that the river flows centrally through the weir constructed across it without damaging the structure and its approaches. The alignment of guide banks should be determined such that the pattern of flow at the head regulator induces favourable curvature of flow minimising sediment entry into the canal system. While constructing weirs on alluvial rivers, the natural waterway is restricted from economic and flow considerations, and the unbridged width of river is blocked by means of approach embankments. Depending upon the distance between the guide bank and the extent of the alluvial belt, the river may form either one or two meander loops. The approach embankments, aligned with the weir axis, should extend up to a point beyond the range of the worst anticipated loop.

Afflux embankments are earthen embankments which extend from both the abutments (or the approach embankments) and are connected on the upstream to the ground above the affluxed highest flood level (or to flood embankments, if existing).

Objectives of River Training

River training measures aim at achieving one or more of the following objectives:

1. Flood Protection

River floods of very small frequency inundate the fertile and thickly-populated plains adjacent to the river, and, thus, cause considerable loss to human life, property, agriculture, and public and private utilities. Statistics collected by the Central Water and Power Commission for flood damage in India during 1953-63 have been given in Table to show the extent of loss due to floods.

Table Average annual flood losses in India during 1953-63 (11):

Population affected	**11.68 million**
Total area affected	64.4 million hectar
Cropped area affected	20.2 million hectares
Human lives lost	518
Total damage to crops and utilities	Rs. 684.0 million

During the years of large floods, damage is likely to be several times more. Flood control measures for thickly-populated flood plains, therefore, become essential, even if these measures do not assure complete protection under all conditions. River training for flood protection, also known as 'high water training' or 'training for discharge,' is achieved by one or more of the following four methods:

- Construction of levees or embankments to confine water in a narrower channel.
- Increasing the discharge capacity of natural channels by some means such as straight-ening, widening or deepening.
- Provision of escapes or diversion from the main channel into an auxiliary channel for water in excess of the carrying capacity of the main channel.
- Construction of reservoirs.

2. Navigation

For a river to be navigable, sufficient depth and width required for navigation should be available even at low water level in the river. River training for navigation is also known as 'low water training' or 'training for depth'. Measures to achieve adequate depth in a river for navigation include dredging the shallow reaches of the river and using spurs to contract the river channel, thus, increasing its depth. Sometimes, low flow is supplemented from another source to achieve the desired depth and width. Canalisation makes a non-navigable river navigable, and, is accomplished by building a series of small dams or weirs and locks. Sharp curves along the river need to be eliminated so that ships can move easily.

3. Sediment Control

River training for sediment control is also called 'mean water training' or 'training for sediment'. This type of training aims at rectification of river bed configuration and efficient movement of sediment load for keeping the channel in a state of equilibrium (3). River training methods for this purpose involve construction of such structures which would induce the desired local curvature to the flow. Spurs and pitched islands are normally used for training the river for sediment.

4. Guiding the Flow

Hydraulic structures, such as canal headworks, and communication structures such as bridges, have to be protected against outflanking and the direct attack of flow. This requires training of the river over its considerable reach by building a system of guide banks, known as Bell's guide banks, on one or both sides of the stream at the bridge site. The purpose of these guide banks is to make sure that water flows between the abutments of the bridge. The spacing between these guide banks conforms to the width required for the river to pass the design flood discharge. Similarly, guide banks are provided to guide the flow at the weir site. Marginal bund and lateral spurs guide the flow through the guide banks.

Sometimes the flow in a river needs to be deflected away from a bank in order to protect some portions of the river bank or for contracting the river. This is done by constructing one or more spurs projecting into the river from its banks.

5. Stabilisation of River Channel

Weak river banks, which are likely to cave in or get eroded, need to be protected by training methods, such as stone pitching, lining, and so on. In some cases, the stability of the bed may also be endangered in some reaches due to increase in the bed shear on account of local flow conditions.

River Training Methods

The planning and design of river training structures is accomplished by using empirical methods and reliance has to be placed on the intuition and judgement of experienced engineers. Model investigations are also resorted to for finalising the plans and design of river training structures. Commonly used methods of river training have been briefly described in the following sections.

Levees

A levee (also known as an embankment, bund, marginal bunds or dike) is an embankment running parallel (or nearly so) to the river and is constructed to protect the area on one side of it from flooding. The method of constructing levees on one or both sides of a river to contain the flood within the leveed portion is the oldest and most commonly used method of flood control. Levees along the Nile River in Egypt were constructed prior to 600 BC. Levees have been constructed recently on many important rivers of the world such as the Ganga, the Kosi, the Mahanadi and the Gandak in India, the Yellow, the Pearl, the Yangtze and the Huai in China, the Mississippi in the USA, and the Danube and the Rhine in Europe.

The alignment of levees for a river is decided by the location of important cities, industries, and other areas along the river which need to be protected against floods. Closely-spaced levees will be very high and, hence, massive and uneconomic.

Chapter 4

Diversion Head Works

4.1 Types of Diversion Head Works

Diversion Head Works

It is constructed at the head of canal, and diverts the river water towards the canal to ensure the regulated and continuous supply of the silt-free water with certain minimum head into the canal.

Objective of these Diversion Head Works:

- To rise the water level at the head of the canal.
- To form the storage by constructing dykes on both the river banks such that water is available throughout the whole year.
- To control the entry of silt into canal and also to control the deposition of the silt at the head of the canal.
- To control the fluctuation of the water level in the river during the different seasons of the year.

Selection of the Site for Diversion Head Works

At the site, the river must be straight and narrow, the river banks must be well defined. The valuable land must not be submerged when the weir or barrage is constructed. Elevation of the site must be much higher than the area that is to be irrigated.

The site must be easily accessible either by roads or railways. The materials of construction must be available in vicinity of the site. The site must not be far away from the command area of the project, in order to avoid transmission loss.

Layout of a Diversion Head Works and its Components

A typical layout of a canal head-works is shown in figure. This head-works consists of:

- Weir proper.
- Under-sluices.

- Divide wall River.
- Training works.
- Fish Ladder.
- Canal Head Regulator.

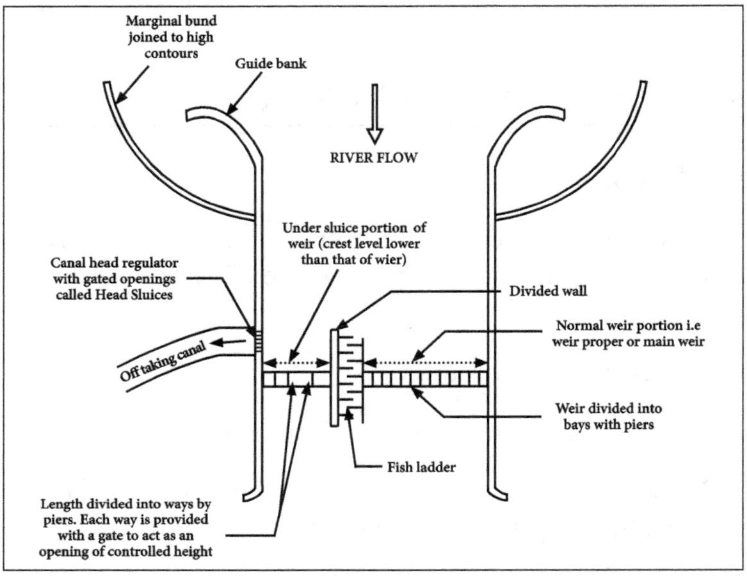

Typical layout of diversion head works.

Weir Proper: It is a barrier built across the river. Its aim is to raise the water level to feed the canal.

Under-sluices: The under sluices are the openings given at the base of weir or barrage. These openings are provided with the adjustable gates. Usually, the gates are kept closed. The crest of under-under sluice portion of the weir is kept at a lower level than the crest of the normal portion of the weir. The suspended silt goes on depositing in front of canal head regulator. When silt deposition becomes appreciable the gates are opened and the deposited silt is loosened with an agitator mounting on a boat. The muddy water flows towards downstream through the scouring sluices. The gates are then closed. Whereas, at the period of flood, the gates are kept opened. The main functions of under-sluices are:

- To maintain a well-defined deep channel approaching the canal head regulator.
- To ensure easy diversion of water into the canal through the canal head regulator even during low flow.
- To control silt entry into the canal.
- To control the entry of silt into the canal.

- To help the scouring and the silt deposited over the under-sluice floor. To also help removing it towards the downstream side.
- To help passing the low floods without dropping the shutters of weir.

The Divide Wall: The divide wall is the masonry or concrete wall constructed at the right angle to the axis of the weir. The divide wall is extending on the upstream side beyond the beginning of canal head regulator and also on the downstream side. It extends upto the end of the loose protection of the under-sluices.

The divide wall is a really long wall built at the right angles in weir or barrage, it can be constructed with the stone masonry or the cement concrete. On the upstream side, the wall is extended just to cover the canal head regulator and on the downstream side, it is extended up to launching apron.

The Main Functions of Divide Walls:

- It separates the 'under-sluices' with the lower crest level from the 'weir proper' with the higher crest level.
- It helps in providing a comparatively less turbulent pocket near canal head regulator, resulting in the deposition of silt in this pocket and, hence, to help in the silt-free water to enter into the canal.
- It helps to keep cross-current away from the weir.

Fish Ladder:

- It is a device by which energy flow can be dissipated in a manner so as to provide smooth flow at essentially low velocity, not exceeding from 3 to 3.5 m/s.
- A narrow opening that included suitable baffles or staggering devices in it is given adjacent to the divide wall.
- The fish ladder is provided just by the side of divide wall for the free movement of the fish. Rivers are important source of fish.
- There are several types of fish in the river. The nature of each fish varies from type to type. But in general, the tendency of fish is to move from the upstream to the downstream in winters and from the downstream to the upstream in monsoons.
- This movement is very essential for their survival. Due to the construction of weir or barrage, this movement gets obstructed, and they are detrimental to the fishes.
- In the fish ladder, fable walls are constructed in a zigzag manner so that the velocity of flow within ladder does not exceed 3 m/sec.

- The width, length and the height of the fish ladder depends on the nature of the river and the type of the weir or barrage.

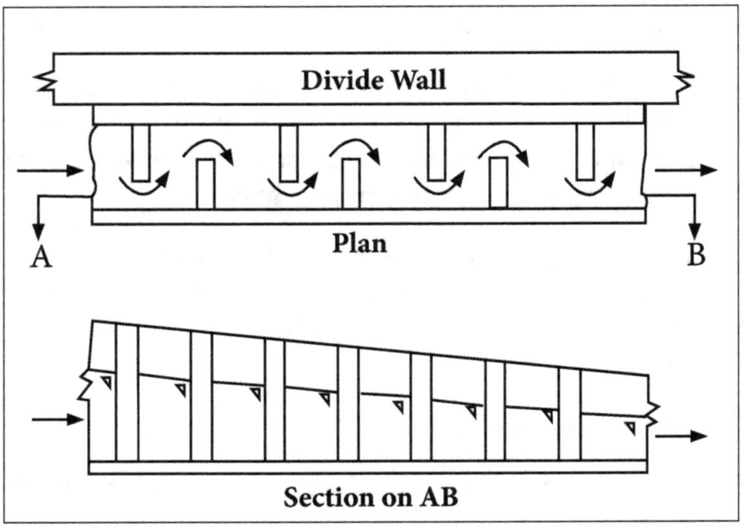

Canal Head Regulator or Head Sluices: The structure which is constructed at the head of the canal in order to regulate flow of water is called as canal head regulator. It has a number of piers which divide the total width of the canal into number of spans which are termed as bays. The piers consist of number tiers on which adjustable gates are present. The gates are operated from the top by a suitable mechanical device. A platform is provided on the top of the piers for facility of gate operation. Again some piers are constructed on downstream side of canal head to support the roadway.

Head Regulator:

- It regulates the water supply that enters into the canal.
- It controls the entry of silt in the canal.
- It prevents river-floods from entering into canal.

The water from the under-sluice pocket is made to enter regulator bays, such that to pass the full supply discharge into canal. The maximum height of such gated openings, known as head sluices will be equal to difference of the Pond Level and the Crest Level of the regulator:

- The entry of silt into canal is controlled by keeping crest of head regulator by about 1.2 to 1.5 meters higher than the crest of under-sluices.
- When a silt-excluder is provided, the regulator crest is moreover raised by about 0.6 to 0.7 meter.
- Silt gets deposited in the pocket, and only clear water enters into regulator bays.

- The deposited silt may be easily scoured out regularly, and removed through under-sluice openings.

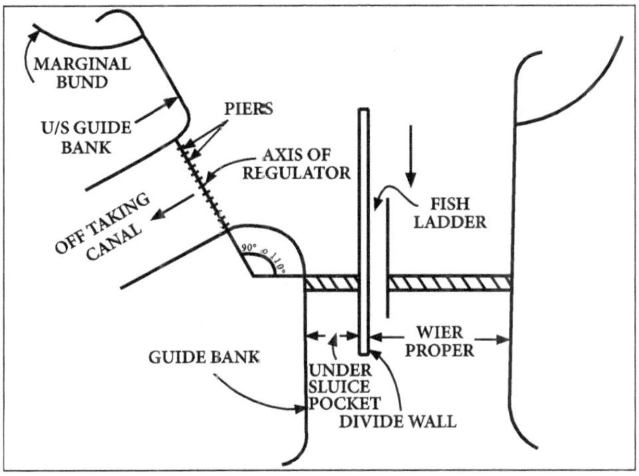

Alignment of a canal head regulator.

Weir and Barrage

It is the barrier built across river to raise the water level on the upstream side of obstruction to feed the main canal.

The ponding of the water may be achieved either only by a raised crest across the river or by the raised crest supplemented by gates or the shutters, working over crest.

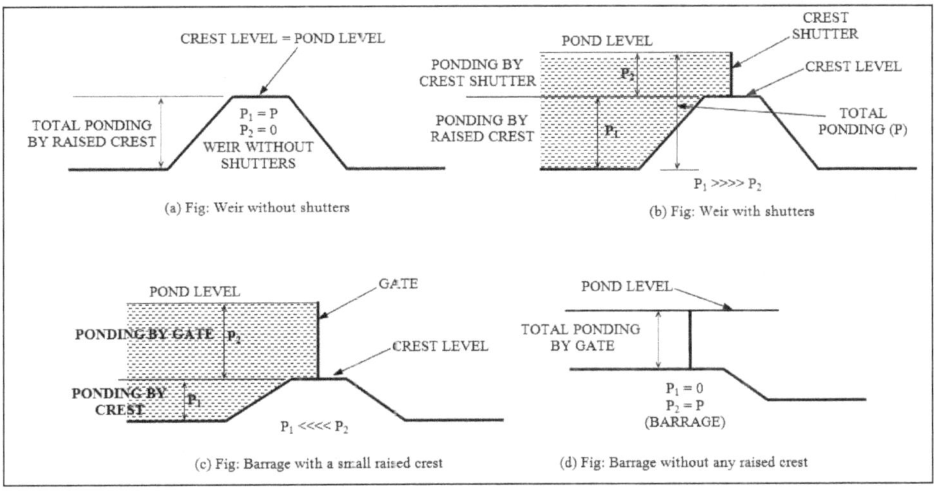

(a) Fig: Weir without shutters

(b) Fig: Weir with shutters

(c) Fig: Barrage with a small raised crest

(d) Fig: Barrage without any raised crest

Weir

If major part or entire ponding of the water is achieved by a raised crest and a smaller part or nil part of it, is achieved by means of shutters, then this barrier is called as a weir.

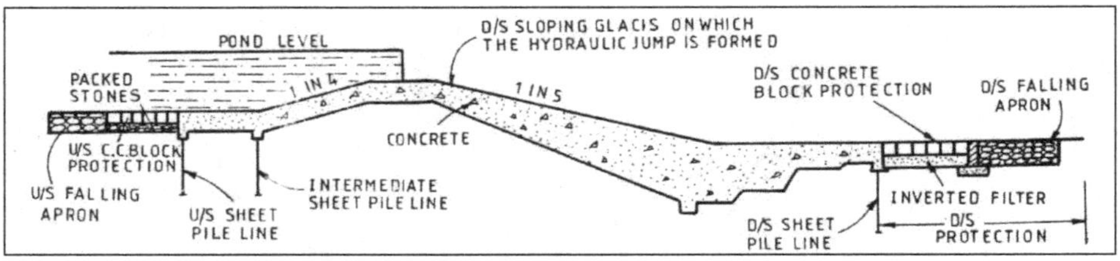

A typical cross-section of a modern concrete weir.

Gravity and Non-gravity Weirs

When the weight of weir balances the uplift pressure caused by head of the water seeping below weir, it is known as a gravity weir. On the other hand, if the weir floor is designed continuous with the divide piers as reinforced structure, in a manner that the weight of concrete slab together with weight of divide piers keeping the structure safe against the uplift ,then structure can be called as a non-gravity weir.

- In the latter case, RCC is to be used in place of the brick piers.
- Considerable savings can be obtained, as weight of the floor may be much lesser than what is needed in gravity weir.

Types of Weirs

- Masonry weirs with vertical drop.
- Rock-fill weirs with sloping aprons.
- Concrete weirs with sloping glacis.

Masonry Weirs with Vertical Drop: Masonry weir wall is constructed over impervious floor. Cut-off walls are present at both ends of the floor. Sheet piles are provided below the cut off walls. The crest shutters are provided to raise water level, if needed. The shutters are dropped down during the flood. The masonry weir wall can be vertical on both face or sloping on both face or vertical on the downstream face and sloping in the upstream face.

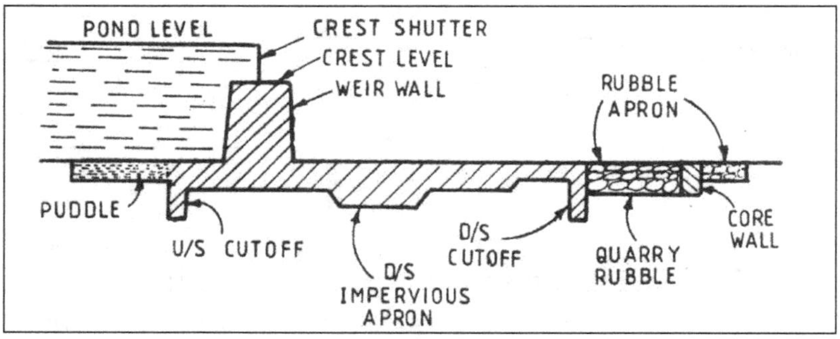

Masonry weir.

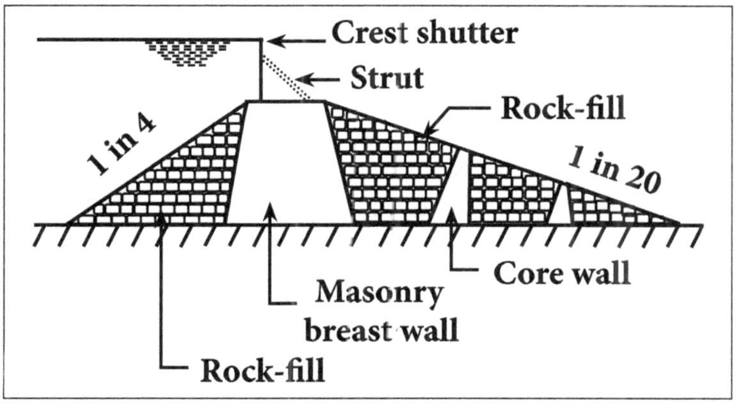

Rock fill wier.

Rock-fill Weirs with Sloping Aprons: It consists of a masonry breast wall which is provided with the adjustable crest shutter. The upstream rock-fill portion is constructed with the boulders forming a slope of 1 in 4. The boulders are grouted with the cement mortar. The downstream sloping apron consists of the core walls. The intermediate spaces between the core walls are filled up with the boulders maintaining a slope of 1 in 20. The boulders are grouted properly with the cement mortar.

Concrete Weir:

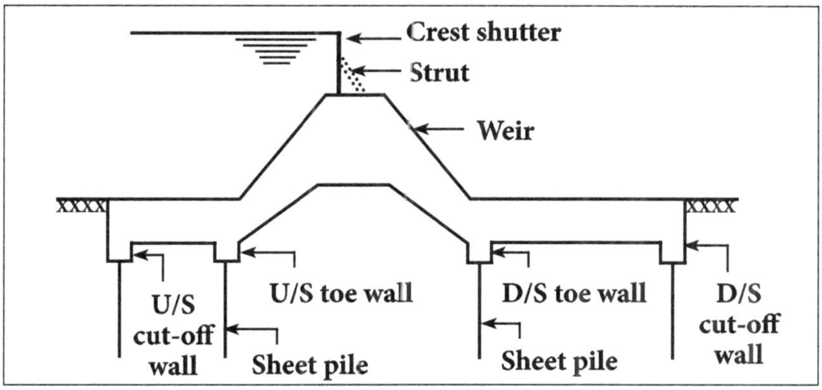

Concrete wier.

Now-a-days, the weirs are constructed with the help of reinforced cement concrete. The impervious floor and the weir are both made monolithic. The cut off walls are provided at the upstream and the downstream end of the floor and at the toe of the weir. Sheet piles are provided below the cut-off walls. The crest shutters are provided which have dropped down during flood.

Barrage

If most of the ponding is done by gates and the smaller or nil part of it is done by the raised crest, then the barrier is called as a barrage or a river regulator.

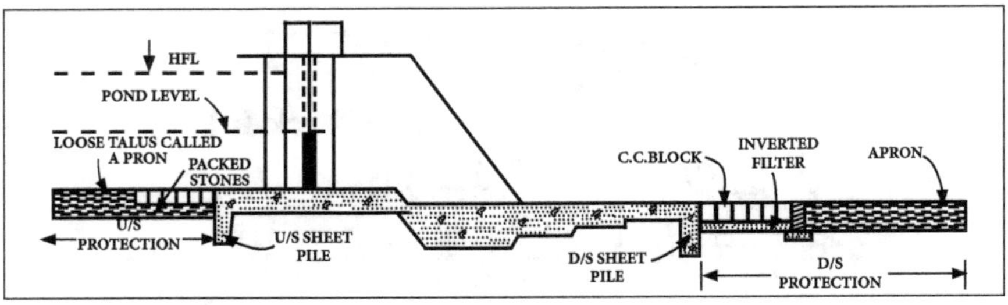

A Typical Cross-section of a Barrage: The only difference between a weir and a barrage is of the gates, that is the flow in which the barrage is regulated by gates and that in weirs, by its crest height.

Barrages are Costlier than Weirs: Weirs and barrages are constructed mostly in the plain areas. The heading up of water is affected by gates that are put across the river. The crest level in the barrage is kept at the low level.

During flood, gates are raised to clear the high flood level. As a result there is less silting and provide better regulation and control than the weir.

Components of Barrage

Main barrage portion: Main body of the barrage, normal RCC slab in which it supports the steel gate. In the X-Section it consists of: Upstream concrete floor, to lengthen the path of the seepage and to project the middle portion where the pier, gates and bridge are located.

A crest at the required height above the floor on which the gates rest in their closed position.

The Upstream glacis of suitable slope and shape. This joins the crest to the downstream floor level. The hydraulic jump forms on the glacis since it is more stable than on the horizontal floor, this eventually reduces length of the concrete work on the downstream side.

Downstream floor is built with the help of concrete and is constructed so as to contain the hydraulic jump. Therefore it takes care of turbulence which would otherwise cause erosion. It is also provided with the friction blocks of suitable shape and at a distance determined through the hydraulic model experiment to increase friction and to destroy the residual kinetic energy.

Divide Wall

A wall constructed at right angle to the axis of the weir separating the weir proper from the under sluices .It extends upstream beyond the beginning of canal HR. Downstream it extends up to the end of loose protection of under sluices.

This is to cover the hydraulic jump and resulting turbulence.

The Fish Ladder

- For movement of fish.
- Difference of level on upstream and downstream sides on weir is split up into water steps by means of the baffle walls that is constructed across the inclined chute of fish ladder.
- Velocity in the chute should not be more than 3m/s.
- Grooved gate at the upstream and the downstream - for effective control.
- Optimum velocity 6-8 ft/s.

Sheet Piles

It is made of mild steel, each portion being 1/2' to 2' in width and 1/2" thick and of the required length, having groove to link with the other sheet piles.

Upstream Piles

It is situated at upstream end of upstream concrete floor driven into soil beyond the maximum possible scour that shall occur.

Functions:

- Protects barrage structure from scour
- Reduces uplift pressure on barrage
- Holds the sand compacted and densified between two sheet piles and thereby increases the bearing capacity when barrage floor is designed as raft.

Intermediate Sheet Piles: It is situated at the end of upstream and downstream glacis. Protection to main structure of barrage in the event of upstream and downstream sheet piles collapsing due to the advancing scour or undermining. They also help lengthen the seepage path and reduce the uplift pressure.

Downstream Sheet Piles: It is placed at the end of the downstream concrete floor. Their main function is to check exit gradient. Their depth must be greater than the possible scour.

Inverted Filter

It is provided between downstream sheet piles and flexible protection. Typically 6" sand, 9" coarse sand and 9" gravel. Filters shall vary with size of the particles forming the river bed. It is protected by placing over the concrete blocks of sufficient weight and size. Slits are left between the blocks to allow the water to escape.

The length must be 2x downstream depth of sheet.

Functions:

- Checks the escape of fine soil particles in the seepage water.
- Flexible apron.
- Placed on the downstream of the filter.
- Consists of boulder large enough not to be washed away by the highest likely velocity.
- The protection provided is enough as to cover the slope of scour of 1 1/2 x depth of scour as the upstream side of 2x depth of the scour on the downstream side at the slope of 3.

The Under Sluices

- Scouring sluices.
- Maintaining a deep channel in front of the Head regulator on the downstream side.

Functions

As the bed of under sluice is not the lower level than rest of the weir, most of the day, whether flow unit will flow toward this pocket => easy diversion to the channel through the Head regulator:

- Control silt entry into channel.
- Scour the silt.
- High velocity currents due to high differential head.
- Passes the low floods without dropping.
- The shutter of the main weir, the raising of which entails good deal of labor and time.

Capacity of the Under Sluices

- For sufficient scouring capacity, its discharging capacity must be at least double the canal discharge.

- Must be able to pass the dry weather flow and the low flood, without dropping the weir shutter.
- It is capable of discharging 10 to 15% of high flood discharge.

Choice between a Weir and a Barrage

The choice between a weir and a barrage is largely governed by the cost and convenience in working:

- A shuttered weir could be relatively cheaper but will lack the effective control possible in the case of a barrage.
- A barrage type construction may be easily supplemented with a roadway across the river at a small additional cost. Barrages are almost invariably constructed now-a-days on all the important rivers.

4.2 Failures of Weirs on Permeable Foundations

Causes of Failure of the Weirs and Their Remedies

The common causes of failure of the weirs includes:

- Progressive and Excessive downstream erosion, both from within stream and through lateral erosion of banks.
- The erosion of inadequately protected abutments.
- Hydraulic removal of fines and other support material from downstream protection resulting in erosion of the apron protection.
- Deterioration of the cutoff and subsequent loss of the containment.
- Additional aspects specific to concrete, rockfill or the steel structures.

The main causes are:

1. Piping

Piping is caused by the groundwater seeping out of the bank face. Grains are detached and are entrained by the seepage flow and can be transported away from the bank face by the surface runoff generated by seepage, if there is sufficient volume of flow.

The exit gradient of water seeping under base of the weir at downstream end shall exceed a certain critical value of the soil. As a result the surface soil starts boiling and they are washed away by the percolating water. The progressive erosion backwash at the upstream results in the formation of channel underneath the floor of weir.

Since there is always a differential head between the upstream & the downstream, water is constantly moving from upstream to downstream from under the base of weir. However, if the hydraulic gradient becomes big, greater than the critical value, then at the point of existence of water at downstream end, it begins to dislodge soil particles and also carry them away.

In due course, when this type of erosion continues, a sort of pipe or channel is formed within the floor through which more particles are being transported downstream which may bring about failure of weir.

Piping is likely in high banks backed by valley side, a terrace, or some other high ground. In these locations the high head of water may cause large seepage pressures to occur. Evidence included are: Pronounced seep lines, especially along sand layers or lenses in bank; pipe shaped cavities in bank; notches in the bank associated with the seepage zones and layers; run-out deposits of eroded material on the lower bank.

Remedies:

- Decrease the Hydraulic gradient
- Providing curtains or piles at both upstream and downstream.

2. Rupture of Floor Due to Uplift

If the weight of floor is insufficient to resist the uplift pressure, then floor shall burst. This bursting of the floor will reduce the effective length of the impervious floor, which will result in increasing the exit gradient, and may cause failure of weir.

Remedies: By Providing impervious floor of sufficient length of appropriate thickness, Pile at upstream to reduce the uplift pressure downstream.

3. Rupture of the Floor Due to the Suction Caused by Standing Waves

Hydraulic jump formed at the downstream of water.

Remedies:

- Additional thickness.
- Floor thickness in one concrete mass.

4. Scour on the Upstream and Downstream of the Weir

Scouring in Weirs - Cause of Failure Scouring in Weirs occurs due to contraction of the natural water way.

Remedies: Piles at a greater depth than scour level.

Launching aprons: Stones of aprons can settle in the scour hole.

4.3 Bligh's Creep Theory, Khosla's Theory, Design of Impervious Floors for Subsurface Flow and Exit Gradient

4.3.1 Bligh's Creep Theory

Flow According to the Bligh's Theory, the percolating water follows the outline of the base of foundation of the hydraulic structure. In other words, water creeps along the bottom contour of structure. The length of the path hence traversed by water is known the length of the creep. Moreover, it is assumed in this theory, that the loss of head is proportional to the length of the creep. If HL is total head loss between the upstream and the downstream, and L is length of creep, then the loss of head per unit of creep length is known as the hydraulic gradient. Further, Bligh makes no distinction between horizontal and vertical creep.

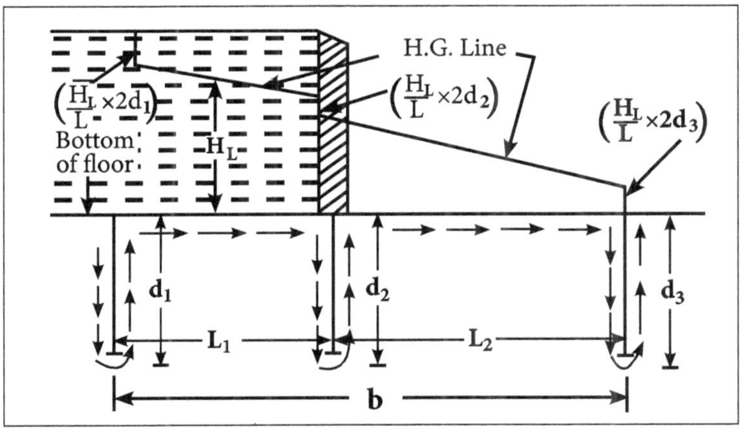

Bligh's creep.

Bligh called the loss of head per unit length of creep as the Percolation coefficient. The reciprocal, (L/H) of the percolation coefficient is called as the coefficient of creep C.

Assumptions

Hydraulic slope or gradient is constant throughout the impervious length of the apron. The percolating water creep along the contact of the base profile of the apron with the sub soil losing head enroute, proportional to the length of its travel. The length is called the creep length. It is the sum of horizontal and vertical creep. Stoppage of the percolation by the cut off possible only if it extends up to the impermeable soil strata.

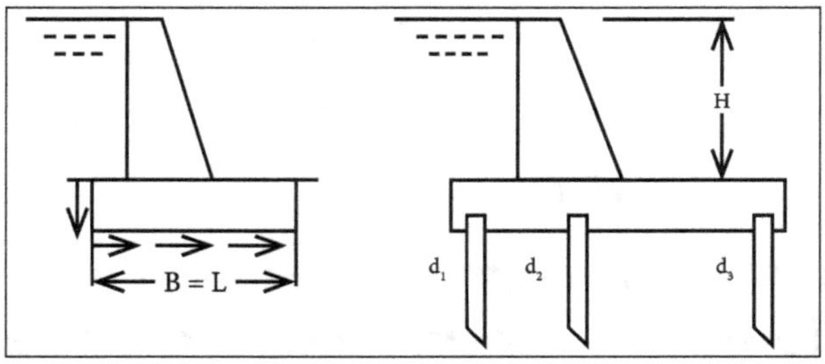

Consider a section a shown in Figure above. Let HL be the difference of water levels between the upstream and the downstream ends. Water will seep along the bottom contour as shown by the arrows. It starts percolating at A and emerges at B. The total length of the creep is given by,

$$L = d_1 + d_1 + L_1 + d_2 + d_2 + L_2 + d_3 + d_3$$
$$= (L_1 + L_2) + 2(d_1 + d_2 + d_3)$$
$$= b + 2(d_1 + d_2 + d_3)$$

Head loss per unit length or gradient,

$$C' = \frac{H}{L} = \frac{H}{[B + 2(d_1 + d_2 + d_3)]}$$

Head losses equal to $\left(\frac{H_L}{L} \times 2d_1\right), \left(\frac{H_L}{L} \times 2d_2\right), \left(\frac{H_L}{L} \times 2d_3\right)$ will occur respectively, in the planes of the three vertical cut offs. The hydraulic gradient line may then be drawn as the shown in figure above.

1. Safety Against Piping or Undermining

According to Bligh, the safety against piping may be ensured by providing the sufficient creep length, given by L = C.HL, where C is the Bligh's Coefficient for the soil. Different values of C for different types of soils are tabulated in Table –1 below:

S. No	Type of soil	Value of C	Safehydraulic gradient should be less than
1	Fine micaceous sand	15	1/15
2	Coarse grained sand	12	1/12
3	Sand mixed with the boulder and gravel, and for loam	5 to 9	1/5 to 1/9
4	Light sand and mud	8	1/8

Note: The hydraulic gradient i.e. HL/L is then equal to 1/C. Hence, it may be stated that the hydraulic gradient should be kept under a safe limit in order to ensure safety against piping.

2. Safety Against Uplift Pressure

The ordinates of H.G line above the bottom of the floor represent residual uplift water head at each point. Say for example, if at any point, ordinate of H.G line above the bottom of the floor is 1 m, then 1 m head of water may act as uplift at that point. If h′ meters is this ordinate, then the water pressure equal to h′ meters will act at this point, and it has to be counterbalanced by the weight of the floor of thickness say t.

Uplift pressure = $\gamma w \times h'$ [where γw is the unit weight of water].

Downward pressure = $(\gamma w \times G).t$ [Where G is specific gravity of the floor material].

For equilibrium, $\gamma w \times h' = \gamma w \times G.\ T$,

$$h' = G \times t$$

For equilibrium,

$$\gamma_w \times h' = \gamma_w \times G.t$$
$$\Rightarrow h' = G \times t$$

Subtracting t on both sides, we get,

$$(h'-t) = (G \times t - t) = t(G-t)$$
$$\Rightarrow t = \left(\frac{h'-t}{G-t}\right) = \left(\frac{h}{G-t}\right)$$

Where, $h' - t = h$ = Ordinate of the H.G line above the top of the floor $G - 1$ = Submerged specific gravity of the floor material.

Design of Impervious Floor for Sub Surface Flow

It is directly depended on the possibilities of percolation in porous soil on which the floor is built. Water from the upstream percolates and creeps slowly through the weir base and the subsoil below it. The head lost by the creeping water is proportional to the distance it travels along the base of the weir profile. The creep length should be made as big as possible so as to prevent the piping action. This shall be achieved by providing the deep vertical cut-offs or sheet piles.

Limitations of the Bligh's Theory

This theory made no distinction between the horizontal and the vertical creep:

- It did not explain the idea of exit gradient - safety against the undermining cannot be simply obtained by considering a flat average gradient but by keeping this gradient will be low critical.

- No distinction between the outer and the inner faces of the sheet piles or the intermediate sheet piles, whereas from investigation it is clear, that outer faces of the end sheet piles are much effective than the inner ones.
- Losses of head does not take place in the same proportions as creep length.

The uplift pressure distribution is not linear but it follows a sine curve.

In case of two piles the width between must be greater than twice the head or the piles are not effective.

4.3.2 Khosla's Theory

The main principles of this theory are summarized below: (a) The seepage water does not creep along bottom contour of pucca flood as started by the Bligh, but on the other hand, this water moves along the set of stream-lines. This steady seepage in a vertical plane for a homogeneous soil may be expressed by Laplacian equation,

$$\frac{d^2\phi}{dx^2}+\frac{d^2\phi}{dz^2}$$

Where, φ = Flow potential = Kh; K = the co-efficient of permeability of soil as defined by Darcy's law, and h is the residual head at any point within the soil.

The above equation represents sets of two individual curves intersecting each other orthogonally. The resultant flow diagram showing both of the curves is called a Flow Net.

Stream Lines

The streamlines represent the paths along which the water flows through the sub-soil. Every particle entering the soil at a given point upstream of the work, will trace out its own path and may represent a streamline. The first streamline follows the bottom contour of the works and is same as Bligh's path of creep. The remaining streamlines follows smooth curves transiting slowly from outline of the foundation to a semi-ellipse, as shown:

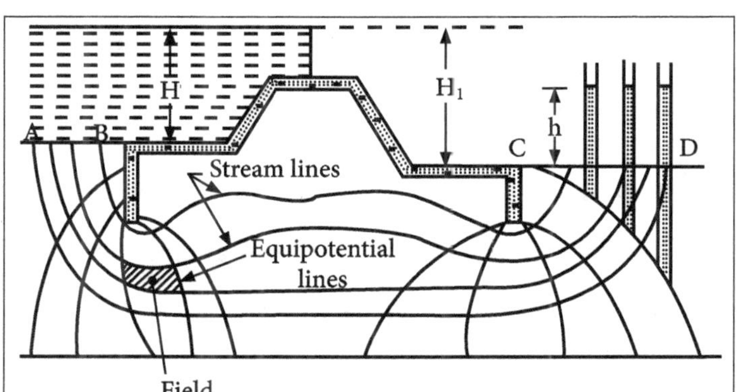

Khosla's flow net.

Equipotential Lines

It treats the downstream bed as datum and assuming no water on the downstream side, it may be easily started that every streamline possesses a head similar to h1 while entering into the soil; and when it emerges at the down-stream end into the atmosphere, its head is zero. Therefore, the head h1 is entirely lost during the passage of water along the streamlines. Further, at every intermediate point in its path, there is a certain residual head (h) still to be dissipated in remaining length to be traversed to downstream end. This fact is applicable to every streamline, and thus, there will be points on different streamlines having same value of residual head h. If such points are joined together, the curve obtained is known as an equipotential line.

Every water particle on line AB having a residual head $h = h_1$, and on CD having a residual head h= 0, and thus, AB and CD are the equipotential lines. As an equipotential line represents the joining of points of equal residual head, thus if piezometers were installed on an equipotential line, the water will rise in all of them up to same level as shown in figure below:

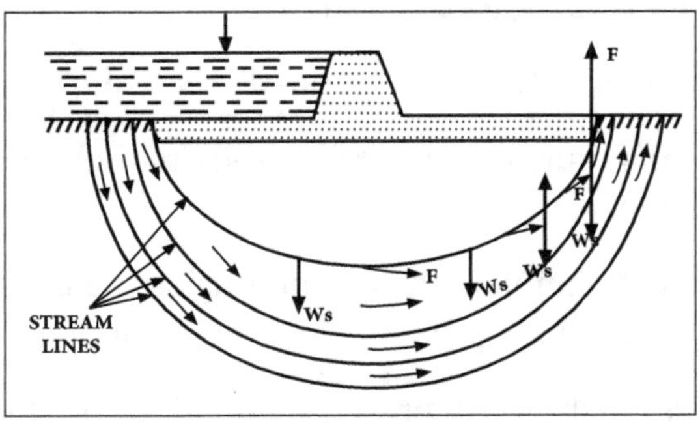

The seepage water exerts a force at each point in the direction of the flow and tangential to the streamlines as shown in figure above. This force (F) has an upward component from the point where streamlines turns upward. For soil grains to remain stable, the upward component of this force must be counterbalanced by the submerged weight of the soil grain. This force has the maximum disturbing tendency at the exit end, this is because the direction of this force at the exit point is vertically upward, and therefore full force acts as its upward component. For soil grain to remain stable, the submerged weight of the soil grain must be more than this upward disturbing force. The disturbing force at any point is always proportional to the gradient of pressure of water at that point. This gradient of pressure of water at the exit end is known as the exit gradient. In order that the soil particles at exit to remain stable, the upward pressure at exit must be safe. In other words, the exit gradient must be safe.

Critical Exit Gradient

This exit gradient is said to be critical, when upward disturbing force on the grain is just equal to the submerged weight of the grain at the exit. When a factor of safety equal to

the 4 to 5 is taken into account, the exit gradient may then be taken as safe. In other words, an exit gradient that equals to ¼ to 1/5 of the critical exit gradient is ensured, so as to keep the structure safe against piping. The submerged weight (W_s) of a unit volume of the soil is given as,

$$\gamma_w (1-n)(S_s - 1)$$

Where,

γ_w = unit weight of water.

S_s = Specific gravity of the soil particles.

n = Porosity of the soil material.

For some critical conditions to occur at the exit point,

$$F = W_s$$

Where F is the upward disturbing force on the grain Force,

F = pressure gradient at that point = $dp/dl = \gamma_w \times dh/dl$.

After studying a lot of dam failure constructed based on Bligh's theory, Khosla came out with the following outcome:

- From observation of Siphons designed on Bligh's theory, by actual measurement of pressure, with the help of pipes inserted in the floor of two of the siphons.

It does not show any relationship with pressure calculated on Bligh's theory. This led to the following provisional conclusions:

- The outer faces of end sheet piles were much more effective than the inner ones and the horizontal length of floor.
- The intermediated piles of smaller length were ineffective except for local redistribution of pressure.
- Undermining of floor started from tail end.
- It was essential to have a reasonably deep vertical cut off at downstream end to prevent undermining.
- Khosla and his associates took into account the flow pattern below the impermeable base of hydraulic structure to calculate uplift pressure and exit gradient.
- Starting with a simple case of horizontal flow with negligibly small thickness.

- Seeping water below a hydraulic structure does not follow the bottom profile of the impervious floor as stated by Bligh but each particle traces the path along a series of streamlines.

For the case of two dimensional flows under the straight floor where:

- Hence for first flow line AB which touches outline of the floor, pressure shall be determined by putting the different values of x in the equation. Figure shows the pressure distribution diameter both by the equation 4 as well as the Bligh's Theory.

From the figure the following conclusions may be drawn:

- Slope of Pressure diagram: Having A and B in infinite, thus the floor at A will be theoretically infinite acting downward and that at B may also be infinite acting upward. This will cause sand boiling and so the floor should be depressed or cut off must be provided at downstream end.

Composite Profile

The following specific causes of the general form were considered:

- Straight and horizontal floor of negligible thickness with pile at the either end, upstream or at the downstream end.
- Straight and horizontal floor of negligible thickness with pile at some intermediate point.
- Straight and horizontal floor, depressed below the bed, but with no cut off.

Method of Independent Variable

- Most designs do not confirm to elementary profiles. In actual cases we may have a number of piles at upstream level, downstream level and the intermediate points and the floor may also have some thickness.
- Khosla solves the actual problem by using an empirical method called as the method of independent variables.
- This method consists of breaking up complex profile into a number of simple profiles, each is independently amiable to the mathematical treatment. Then by applying corrections due to the thickness of slope of floor.
- As an example complex profile shown in figure is broken up to the following simple profile and the pressure at Key Points are calculated.
- Straight floor of negligible thickness with pile at the upstream ends.
- Straight floor of negligible thickness with pile at the downstream end.

- Straight floor of negligible thickness with pile at the intermediate points.
- The pressure is obtained at the key points by considering the simple profile.

For the determination of seepage below the foundation of the hydraulic structure developed the method of independent variable. In this method, actual profile of a weir which is complex, is divided into a number simple profiles, each of which can be solved mathematically without difficulty. The most useful profile considered are:

A straight horizontal floor of negligible thickness provided with a sheet pile at the upstream end or a sheet pile at the downstream end.

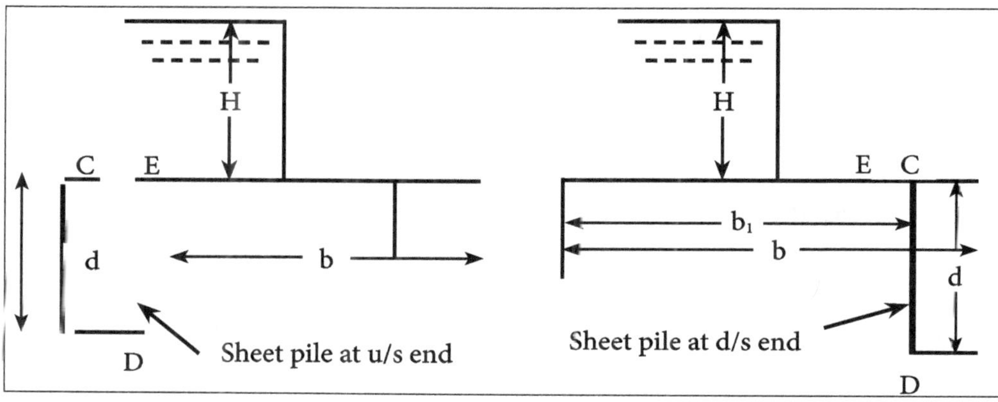

- A straight horizontal floor depressed below the bed, but without any vertical cut-off.
- A straight horizontal floor of negligible thickness with a sheet pile at some.

Intermediate Point

The mathematical solution of the flow-nets of the above profiles have been given in the form of curves. From the curves, percentage pressures at various key points can be determined. The important points to note are:

- Junctions of pile with the floor on either side.
- Bottom point of pile (D).
- Junction of the bottom corners (D, D') in case of the depressed floor.

The percentage pressures at the key points of a simple forms will become valid for any complex profile, provided the following corrections are effected:

- Corrections for mutual interference of piles.
- Corrections for the thickness of floor.

- Corrections for slope of the floor.

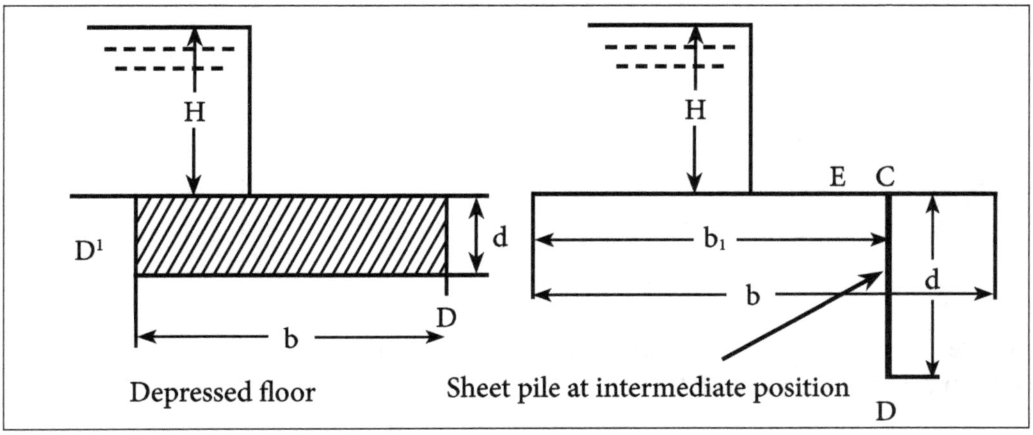

Corrections for Mutual Interference of Piles

Let,

b_1 = distance between the two piles 1 and 2.

D = the depth of the pile line (2), the influence of which on the neighbouring pile (1) of depth d should be determined.

b = total length of impervious floor.

c = corrections due to the interference.

The correction is applied as the percentage of the head,

$$C = 19\sqrt{\frac{D}{b_1}} x \left(\frac{d+D}{b}\right)$$

This correction is positive if the point is considered to be at the rear of interfering pile and negative for the points considered in the forward or flow of direction with the interfering pile.

Corrections for the Floor Thickness

Standard profiles assuming the floors as having negligible thickness. Hence the values of the percentage pressures computed from the curves corresponds to the top levels (E_1^*, C_1^*) of the floor. Moreover, the junction points of the floor and pile are at the bottom of the floor (E_1, C_1).

The pressures at an actual points E_1 and C_1 are interpolated by assuming a straight line variation in the pressures from the points E_1^* to D_1 and from D_1 to C_1.

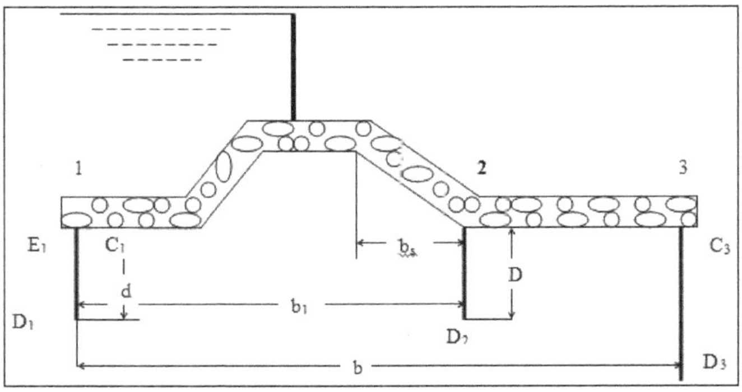

The corrected pressures at E_1 must be less than computed pressure tE_1^*. Hence the correction for the pressure at E_1 will be negative. And so also is for pressure at C_1.

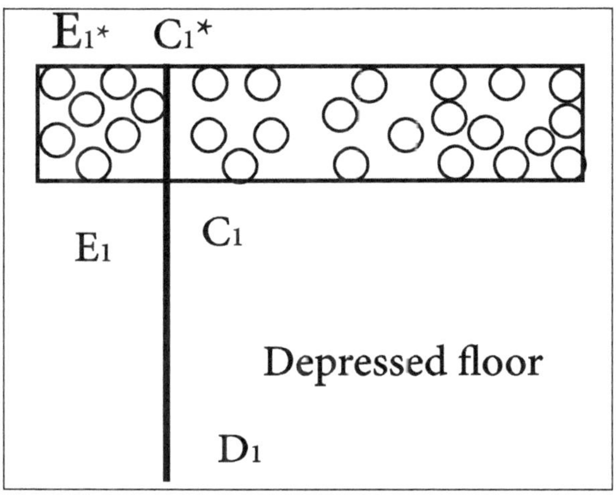

Correction for Slope of Floor

A correction for sloping impervious floor is positive for down slope in the flow direction and negative for the up slope in the direction of flow:

No.	Slope = Ver: Horiz	Correction as % of pressure
1	1:1	11.2
2	1:2	6.5
3	1:3	4.5
4	1:4	3.3
5	1:5	2.8
6	1:6	2.5
7	1:7	2.3
8	1:8	2.0

The correction factor should be multiplied by the horizontal length of the slope and divided by the distance between two poles between which the sloping floor exists.

In the diagram above, correction for slope can be applied only to point E_2. As the point E_2 is terminating at descending slope in the direction of flow, the correction will be positive. The value of correction will be,

$$C.F. \times b_s/b_1$$

Where,

C.F. = is the correction factor.

b_s = horizontal length of the sloping floor.

b_1 = horizontal distance between pile lines.

Exit Gradient

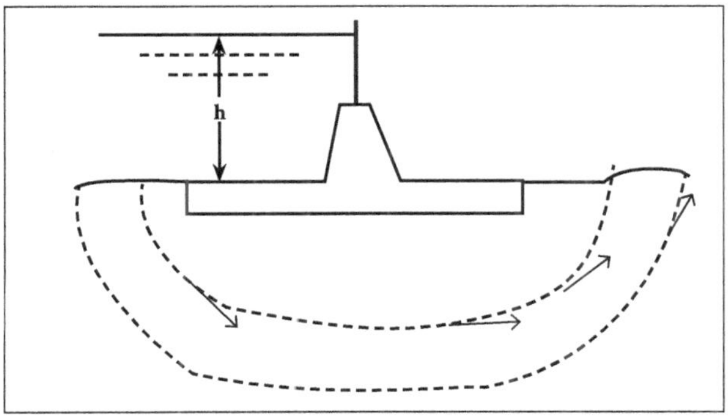

Exit gradient.

Every particle of water when seeping through the sub-soil, at any position may exert a force f, which may be tangential to streamline at any point. As the streamlines bend upward, the tangential force f will be having a vertical component f1.At that point, there will be a downward force W due to the submerged weight of the soil particle. Thus at that point there will be two forces on particle where one is upward vertical component of f, and the other, is the submerged weight. It is evident that if the soil particle is not to be dislodged, then the submerged weight should be greater than upward vertical component of f. The upward vertical component force at any point is always proportional to the water pressure gradient dp/dx.

Thus for stability of the soil and for the prevention of erosion and piping, the seeping water when it emerges at the downstream side, at the exit position, the force f1 must be less than the submerged weight W. In other words the exit gradient at the downstream end should be safe.

If at the exit point at the downstream side, the exit gradient is such that force f1 is equal to the submerged weight of soil particle, then this gradient is known as the Critical gradient. The Safe exit gradients = 0.2 to 0.25 of the critical exit gradient.

Values of the safe exit gradient can be taken as,

0.14 to 0.17 for fine sand

0.17 to 0.20 for coarse sand

0.20 to 0.25 for shingle

For the standard form consisting of floor of length b, and vertical cut-off of depth d, the exit gradient at its downstream end is given as,

$$\text{Exit gradient GE} = (H/d) \times \frac{1}{\pi\sqrt{\gamma}}$$

Where,

$$\lambda = \frac{1+\sqrt{1+\alpha^2}}{2}$$

$$\alpha = b/d$$

H = Maximum Seepage Head.

Chapter 5

Reservoir Planning and Dams

5.1 Reservoir Planning

A reservoir is a large artificial lake created by building a dam across a river. Broadly speaking, any water pool or a lake can be termed as a reservoir. However the term reservoir in the context of water resources engineering is used in a restricted sense for comparatively large body of water stored on upstream of a dam constructed for this purpose. Thus a reservoir and a dam exist together. The discharge in a river generally varies considerably during different periods of the year.

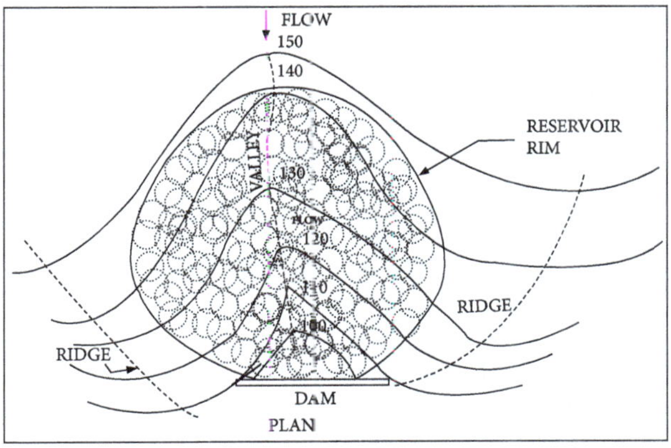

Reservoir

If a reservoir is for serving only one purpose, it is known as a single-purpose reservoir. On the other hand if it serves more than one purpose, it is known as multipurpose reservoir. The various purposes served by a multipurpose reservoir include:

- Municipal and industrial water supply.
- Irrigation.
- Hydropower.
- Flood control.
- Recreation.
- Navigation.

- Soil conservation.
- Development of fish and wild life.
- Pollution control.
- Mosquito control.

Investigations

The basic data, usually required for planning of dams and reservoirs, can be grouped in the following categories:

- Hydrologic data: Stream flows, flood flows, evaporation, sedimentation, water quality, water rights, and tail-water curves.
- Geological data: Reservoir sites, dam sites, and construction materials.
- Topographic surveys: Catchments, reservoir sites, dam sites, and borrow areas.
- Legal data: Water rights.
- Reservoir site cost data: Land acquisition, clearing, and relocations.
- Environmental factors: Fish and wildlife, recreation, scenic, historical, and archaeological.
- Economic data: Economic base for area benefited, crop data, land classification, and market data for various purposes.

The desirable quality of these data would depend on the level of investigations. Investigations for a water resource project are generally carried out in three separate steps (or levels or stages): reconnaissance (or preliminary), feasibility, and pre-construction.

5.1.1 Zones of Storage Reservoirs

Normal Pool Level or Maximum Conservation Level

It is the maxi-mum elevation to which the reservoir water surface will rise during normal operating conditions. It is equivalent to the elevation of the spillway crest or the top of the spillway gates, for most of the cases.

Minimum Pool Level

The lowest water surface elevation, which has to be kept under normal operating conditions in a reservoir, is called the minimum pool level. This level may be fixed by the elevation of the lowest outlet in the dam or may be guided by the minimum head required for efficient functioning of turbines.

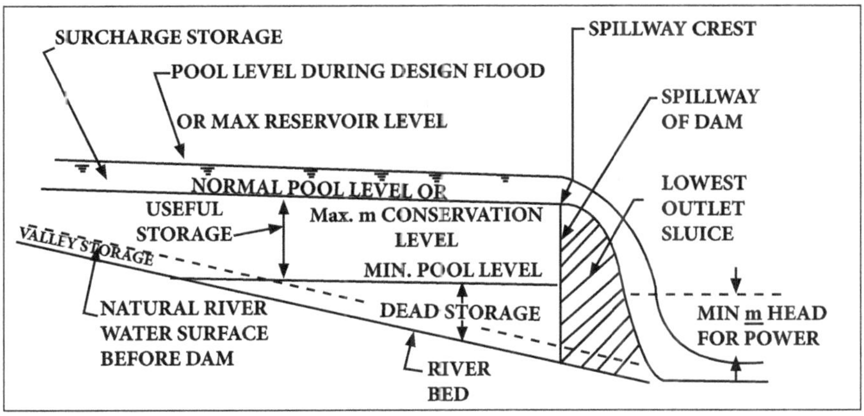

Useful and Dead Storage

The volume of water stored in a reservoir between the minimum pool and normal pool levels is known as the useful storage. Water stored in the reservoir below the minimum pool level is known as the Dead Storage, and it is not of much use in the operation of the reservoirs. The useful storage may be subdivided into conservation storage and flood-mitigation storage, in a multipurpose reservoir.

Maximum Pool Level or Full Reservoir Level

During high floods, water is discharged over the spillway, but will cause the water level to rise in the reservoir above the normal pool level. The maximum level to which the water rises during the worst design flood is known as the maximum pool level.

Surcharge Storage: The volume of water stored between the normal pool level and the maximum pool level is called surcharge storage. Surcharge storage is an uncontrolled storage, in the sense that it exists only till the flood is in progress and cannot be retained for later use.

Bank Storage: When the reservoir is filled up, certain amount of water seeps into the permeable reservoir banks. This water comes out as soon as the reservoir gets depleted. This volume of water is known as the bank storage, and may amount to several percent of the reservoir volume depending upon the geological formations. The bank storage effectively increases the capacity of the reservoir above that indicated by the elevation capacity curve of the reservoir.

Valley Storage: Even before a dam is constructed, certain variable amount of water is stored in the stream channel, called valley storage. After the reservoir is formed, the storage increases, and the actual net increase in the storage is equal to the storage capacity of the reservoir minus the natural valley storage. The valley storage thus reduces the effective storage capacity of a reservoir. It is not of much importance in conservation reservoirs, but the available storage for flood mitigation is reduced, as given by the following relation:

Effective storage for flood mitigation = Useful Storage + Surcharge Storage — Valley Storage corresponding to the rate of inflow in the reservoir.

Types of Reservoirs

Depending upon the purpose served the reservoirs may be broadly classified into five types:

- Storage (or conservation) reservoirs.
- Flood control reservoirs.
- Multipurpose reservoirs.
- Distribution reservoirs.
- Balancing reservoirs.

Storage Reservoirs: Storage reservoirs also are known as conservation reservoirs as a result of their wont to conserve water. Storage reservoirs area unit created to. Store the water within the time of year and to unharness it later once the stream flow is low store reservoirs area unit typically created for irrigation, the municipal installation and hydropower. Though the storage reservoirs area unit created for storing water for varied functions incidentally they conjointly facilitate in tempering the floods and reducing the flood injury to some extent on the downstream. However they're not designed as control reservoirs.

Flood Control Reservoirs: A control reservoir is made for the aim of control. It protects the areas lying on its downstream aspect from the damages because of flood. However absolute protection from extreme floods isn't economically possible. A control reservoir reduces the flood injury and it's conjointly called the flood-mitigation reservoir. Sometimes it's known as flood protection reservoir. During a control reservoir, the floodwater is discharged downstream until the outflow reaches the safe capability of the channel downstream. Once the discharge exceeds the safe capability. The surplus water is hold on within the reservoir. The hold on water is afterward free once the influx to reservoir decreases.

Care is, however, taken that the discharge within the channel downstream, as well as native influx, doesn't exceed its safe capability. A control reservoir is meant to moderate the flood and to not conserve water. However, incidentally some storage is additionally done throughout the amount of floods. Control reservoirs have comparatively massive sluice-way capability to allow speedy drawdown before or once the prevalence of a flood.

Multipurpose Reservoirs: A useful reservoir is meant and created to serve 2 or a lot of functions. Most of the reservoirs area unit designed as useful reservoirs to store water for irrigation and hydropower, and conjointly to result control.

Distribution Reservoir: A distribution reservoir may be a little storage reservoir to bridge over the height demand of water for municipal installation or irrigation. The distribution reservoir is useful in allowing the pumps to figure at a homogeneous rate. It stores water throughout the amount of lean demand and provides an equivalent throughout the amount of high demand. Because the storage is restricted, it just helps in distribution of water as per demand for daily more or less and not for storing it for a protracted amount. Water is pumped-up from a water supply at a homogeneous rate throughout the day for twenty-four hours however the demand varies from time to time. throughout the amount once the demand of water is a smaller amount than the pumping rate, the water is hold on within the distribution reservoir. On the opposite hand, once the demand of water is over the pumping rate, the distribution reservoir is used for supply water at rates larger than the pumping rate. Distribution reservoirs area unit seldom used for the availability of water for irrigation. These area unit in the main used for municipal installation.

Balancing Reservoir: A reconciliation reservoir may be a little reservoir created d/s of the most reservoir for holding water free from the most reservoir.

5.1.2 Yield and Storage Capacity of a Reservoir

Whatever is also the utilization of a reservoir, its most vital operate is to store water throughout floods and to unharness it later. The storage capability of a reservoir is, therefore, its most vital characteristics. The accessible storage capability of a reservoir depends upon the topography of the location and therefore the height of dam to work out the accessible storage capability of a reservoir up to a precise level of water, engineering surveys area unit typically conducted. For correct determination of the capability, a geographics survey of the reservoir space is typically conducted, and a relief map of the world is ready. A contour set up of the world is ready to a scale of one cm = 100m or a hundred and fifty with a contour interval of one to three m, relying upon the dimensions of the reservoir. The storage capability and therefore the water unfold space at completely different elevations is determined from the relief map as explained below:

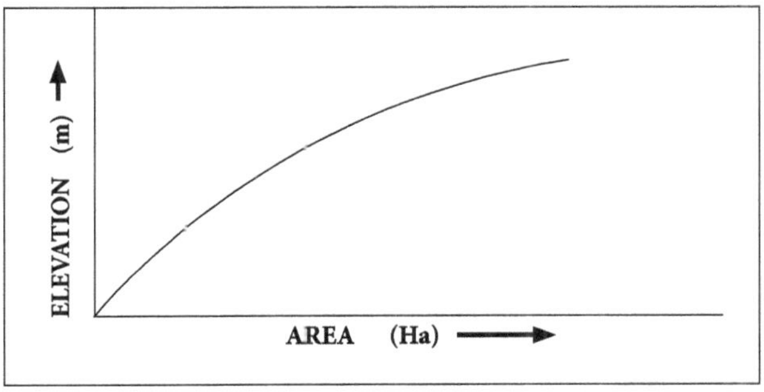

- Area-Elevation Curve: From the contour plan the water spread area of the reservoir at any elevation is determined by measuring the area enclosed by the corresponding contour. Generally a planimeter is used for measuring the area. An elevation-area curve is then drawn between the surface area as abscissa and the elevation as ordinate.

- Elevation-Capacity Curve: The storage capacity of the reservoir at any elevation is determined from the water spread area at various elevations. The following formulae are commonly used to determine the storage capacity (i.e. storage volumes).

Trapezoidal Formula

According to the trapezoidal formula, the storage volume between the two successive contours of areas A_1 and A_2 is given by,

$$\Delta V_1 = \frac{h}{2}(A_1 + A_2)$$

Where h is the contour interval.

Therefore the total volume V of the storage is given by,

$$V = \Delta V_1 + \Delta V_2 + \Delta V_3 + \ldots\ldots\ldots\ldots = \Sigma \Delta V$$

$$V = \frac{h}{2}\left[\Delta V_1 + \Delta V_2 + \Delta V_3 + \ldots\ldots\ldots\ldots + 2A_{n-1} + A_n\right]$$

Where n is the total number of areas.

Cone Formula

According to the cone formula the storage volume between two successive contours of areas A_1 and A_2 is given by,

$$\Delta V_1 = \frac{h}{3}\left(V_1 + V_2 + \sqrt{A_1 A_2}\right)$$

The total volume V is given by,

$$V = \Delta V_1 + \Delta V_2 + \Delta V_3 + \ldots\ldots\ldots\ldots = \Sigma \Delta V$$

Prismoidal Formula

According to the prismoidal formula, the storage volume between 3 successive contours is given as,

$$\Delta V_1 = \frac{h}{3}(A_1 + 4A_2 + A_3)$$

The total volume is given by,

$$V = \frac{h}{3}\left[(A_1 + A_n) + 4(A_2 + A_4 + A_6 + \ldots\ldots) + 2(A_3 + A_5 + \ldots)\right]$$

Where A_3; A_5, etc. are the areas with odd numbers: A_2, A_4, A_6, etc. are the areas with even numbers A_1 and An are respectively the first and the last area.

The prismoidal formula is applicable only when there are odd numbers of areas. In the case of even number of areas the volume up to the last but one area is calculated by the prismoidal formula, and that of the last segment is determined by the trapezoidal formula.

Mass Curve

Of the various methods available for fixing the live storage capacity, the Working Table method may be used which is prepared on the basis of preceding long term data on discharge observation at the site of the proposed reservoir inclusive of at least one drought period. A typical format for carrying out the working table computation is given in the following table:

Months	Begining of period			Inflio w-in M-H-m	Total M-HA-m	Demand during period				End of period			Spilled water	Tail Race Level	Effective Head in m	Discharge in m³/s	Regeneration	Power potential p=7.4 QH
	Reservior level	Capacity in M-Ha-m	Water Spread area			For irrigation	For power	Evaporation	Total	Reservior Level	Capacity M Ha m	Water spread area						
(1)	(2)	(3)	(4)	(5)	(6)= (3) + (5)	(7)	(8)	(9)	(10) = (7) +(8) +(9)	(11)	(12) = (6) - (10)	(13)	(14)	(15)	(16)	(17)	(18)	(19)

The working table calculations may be represented graphically by plotting the cumulative net reservoir inflow exclusive of upstream abstraction as ordinate against time as abscissa. This procedure is commonly called the mass curve technique where the ordinate may be denoted by depth in centimeters or in hectare meters or in any other unit of volume. Discharge with units of 10 days or a month may be used culmination in the mass curve. A segment of the mass curve is shown in Figure:

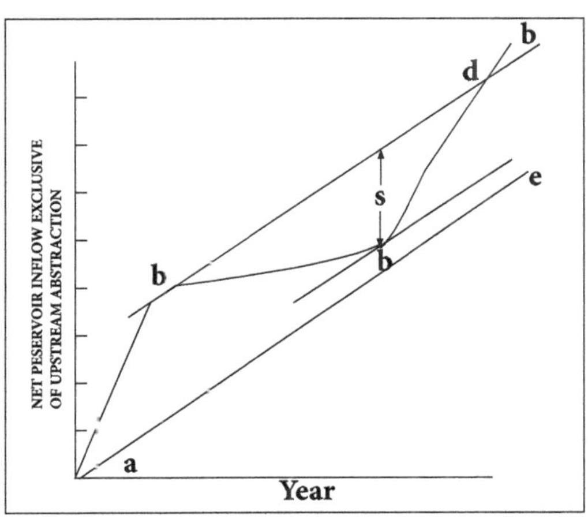

The difference in the ordinate at the end of a segment of the mass curve gives the inflow volume during that time interval. Lines parallel to the lines of uniform rate of demand are drawn at the points b and c of the mass curve. At d the following inferences can be made:

- The inflow rate between a to b is more than the demand rate and the reservoir is full.
- Reservoir is just full as the inflow rate is equal to the demand rate.
- Reservoir storage is drawn down between b and c since demand rate exceeds the inflow rate.
- S is maximum at c due to the demand rate being equal to inflow rate.
- Reservoir is filling or in other words draw down is decreasing from c to d as inflow rate is more than the demand rate.
- Reservoir is full at d and from d to b the reservoir is over flows because the inflow rate exceeds the demand rate. The greatest vertical distance, S at c is the storage required to make up proposed demand.

The withdrawals from the reservoir to meet irrigation demand are basically variable and in these cases the demand line becomes a curve instead of being straight line. The demand mass curve must be super-imposed on inflow mass curve on the same time scale. When inflow and demand mass curves intersect, the reservoir can be assumed to be full. For emptying conditions of the reservoir the demand curve would be above the inflow curve and maximum ordinate between the two indicate the live storage capacity that is required.

Once it is decided to build a reservoir on river by building a dam across it, it is required to arrive at a suitable design capacity of reservoir. The reservoir storage usually consists of three main parts which can be broadly classified as:

- Active or conservation storage.
- Flood and surcharge storage.
- Inactive storage including dead storage.

In general, these storage capacities are to be designed based on certain specified considerations, which have been discussed separately in the following Bureau of Indian Standard codes:

IS: 5477 Fixing the capacities of reservoirs-Methods.

(Part 1): 1999 General requirements.

(Part 2): 1994 Dead storage.

(Part 3): 1969 Line storage.

(Part 4): 1911 Flood storage.

The data and information that is required for fixing the various components of the design capacity of a reservoir are given by the following:

- Precipitation, run-off and the silt records available in region.
- Erodibility of the catchment upstream of the reservoir for estimating the sediment yield.
- Area capacity curves at the proposed location.
- Trap efficiency.
- Losses in the reservoir.
- Water demand from the reservoir for different uses.
- Committed and future upstream uses.
- Criteria for assessing the success of the project.
- Density current aspects and location of outlets.
- Data required for economic analysis.
- Data on engineering and geological aspects.

These aspects are explained in detail in the following sections.

Precipitation, Run-off and Silt Record

The network of precipitation and the discharge measuring stations in the catchment upstream and near the project has to be considered to test the capacity of the same in order to adequately sample both spatially and temporally, along the precipitation and the stream flows.

The measurement procedures and gap filling procedures in respect to the missing data as also any rating tables or curves has to be critically tested so that they are according to instructions of World Meteorological Organization (WMO). Long-term data need to be checked for the internal consistency between rainfall and discharges, as between data sets by double mass analysis to highlight any changes in test data for detection of any long term trends as also for stationarity. It is only after such testing that the data must be used for generating the long term inflows of water into the reservoir.

Sufficiently long term precipitation and run-off records are necessary for preparing the water inflow series. For working out the catchment average sediment yield, long-term data of silt measurement records from existing reservoirs are essential. These are prerequisites for fixing the storage capacity of reservoirs.

If long term run-off records are not available, concurrent rainfall and run-off data shall be used to convert the long term rainfall data into long-term run-off series adopting appropriate statistical/conceptual models. In some cases regression analysis can also be resorted to for data extension.

Estimation of average Sediment Yield from the Catchment Area above the Reservoir

It is generally attempted using river sediment observation data or more commonly from the experience of sedimentation of the existing reservoirs with similar characteristics where observations of stage/flow data is available only for short periods, these have to be suitably extended with help of longer data on rainfall to estimate as far as possible sampling errors due to scanty records. Sediment discharge rating curve can also be prepared from hydraulic considerations using any of standard sediment load formulae, like, modified Einstein's procedure, Young's stream power, etc. It is also required to account for the bed load which is not been measured. Bed load measurement is preferable and it is often estimated as a percentage generally ranging from 5 to 20 percent of the suspended sediment load, when it is not possible,. However actual measurement of bed load has to be undertaken particularly in cases where high bed loads are anticipated.

To assess the volume of the sediment that would deposit in the reservoir, it is moreover necessary to make estimates of average trap efficiency of the reservoir and the likely unit weight of the sediment deposits, along with time average over the period selected. The trap efficiency will depend on the capacity inflow ratio but may also vary with the locations of controlling outlets and reservoir operating procedures. Computations of reservoir trap efficiency can be made using the trap efficiency curves such as those developed by Brune and by Churchill.

Elevation Area Capacity Curves

Topographic survey of a reservoir area should form the basis for obtaining these curves, which are respectively the plots of elevation of the reservoir versus surface area and elevation of the reservoir versus volume. For preliminary studies, in case suitable topographic map with contours, say at intervals less than 2.5 m is not available, stream profile and valley cross sections carried out at regular intervals may form the basis for computing volume. Aerial survey can also be adopted when available facilities are there.

Trap Efficiency

Trap efficiency of any reservoir, over a period, is ratio of total deposited sediment to total sediment inflow. Figures cover relationship between the sedimentation index of the reservoir and the percentage of incoming sediment and these curves can be used for calculation of the trap efficiency.

Losses in Reservoir

Water losses mainly because of evaporation and seepage occur under pre-project conditions and they reflected in the stream flow records which are used for estimating water yield. The construction of new reservoirs and canals is often accompanied by additional evaporation and infiltration. Estimation of those losses can be based on measurements at the existing reservoirs and the canals. The measured inflows and outflows and the rate of change of storage are balanced by the computed total loss rate.

The depth of water evaporated per year from the reservoir surface shall vary from about 400 mm in cool and humid climate to more than 2500 mm in hot and arid regions. Thus, evaporation is an important consideration in many projects and deserves a very careful attention. Several methods such as water budget method, energy budget method, etc. can be applied for calculating the evaporation from reservoir. However, to be accurate, evaporation from reservoir is estimated by using data from the pan-evaporimeters or pans exposed to atmosphere with or without meshing in or near the reservoir site and suitably adjusted.

Seepage losses from the reservoirs and the irrigation canals can be significant when these facilities are located in area underlain by the permeable strata. Avoidance in full or in part of seepage losses will be very expensive and technical difficulties involved will render a project unfeasible. These are basically covered under the conveyance losses in canals projected on the demand side of the simulation studies.

Demand, Supply and Storage

The demand must be compared with the supplies available for every year. When the demand is limited and less than the available run-off, storage can be fixed to cater to that particular demand which is in excess of run-off. The rough and ready method is mass curve method for initial sizing.

While doing the above exercise, water use data are necessary to assess the impact of human activities on the natural hydrological cycle. Sufficient water use information would assist in implementing water supply projects, namely, evaluating the effectiveness of options for demand management and also in resolving problems inherent in competing uses of water, shortages caused by excessive withdrawal, etc. Water demands existing before construction of a water resource project should be considered in the design of project as failure to do so may result in losses apart from legal and social problems at the operation stage.

Committed and Future Upstream uses

The reservoir to be planned must serve not only the present day requirements but the anticipated future requirements. The social, economic and technological developments will bring in considerable difference in the future need or growth rate as compared to the present day need or growth rate. Committed and upstream future uses must also be assessed in the same perspective.

Criteria for Assessing the Success of the Project

Water Resources Projects are to be designed for achieving specified success. Irrigation projects are to be successful for 75 percent period of simulation. Similarly, power projects and water supply projects are to be successful for 90 percent and nearly 100 percent period of the simulation respectively.

Density Current Aspects and Location of Outlets

Density current is defined as the gravitational flow of one fluid under another having slightly different density. The water stored in reservoir is usually free from silt but the inflow during floods is generally muddy. There are, hence two layers that has different densities resulting in the formation of the density currents. The density currents separates the water from the clearer water and make the turbid water flow along river bottom. The reservoir silting rate may be reduced by venting the density currents by properly locating and operating outlets and sluice ways.

Data Required for the Economic Analysis

Economic Analysis is carried out to indicate the economic desirability of the project. The Benefit cost ratio, the Net benefit, the Internal Rate of Return are the parameters in this direction. It is desirable to have the benefit cost ratio in case of irrigation projects and flood mitigation projects to be more than 1.5 and 1.1 respectively. The benefit functions for reservoir and water utilization for irrigation, power, water supply etc., are also to be calculated judiciously. Cost benefit functions are obtained as continuous functions by using variable cost/benefit against reservoir storage/net utilization of water and from benefit functions the benefit from unit utilization of water may be calculated. The spillway capacity has to be adequate to pass the inflow design flood using moderation possible with surcharge storage or any other unobstructed capacity in reservoir without endangering structural safety as provided elsewhere in the standard. In the event of the inflow design flood passing the reservoir, the design needs to be ensured that dam break situation does not develop or induce incremental damage downstream.

Data on Engineering and Geological Aspects

Under engineering and geological aspects the following items of work shall invariably be carried out:

1. Engineering

- Control surveys like topographical surveys.

- Preliminary surveys to assess the catchment and reservoir.
- Detailed survey for making area capacity curves for use in reservoir flood routing.
- Location of nearest Railway lines/Roads and possible access.

2. Geology

- Factors relating to reservoir particularly with reference to water tightness.
- General formations and foundation suitability.
- Deleterious mineral and salt deposits.
- Contributory springs.
- Important aspects of the methods are briefly described in the subsequent sections.
- Location of quarry sites, etc.

Trap Efficiency

Trap efficiency is the proportion of incoming sediment that is deposited or trapped in a reservoir. Most of the estimation methods defines the relationship of the reservoir to their capacity and the annual inflow generally through curves.

5.1.3 Reservoir Sedimentation

The sediment yield from the construction is decided by measure the accumulated sediment during a reservoir for a famous amount by suggests that of echo sounders and alternative electronic devices since the traditional sounding operations provide incorrect ends up in giant depths the amount of sediment accumulated during a reservoir is computed because the distinction between this reservoir capability and also the original capability once the completion of the dam. The unit weight of deposit is decided within the laboratory front the representative undisturbed samples or by field determination employing a mark density probe developed for this purpose.

The total sediment volume is then regenerate to dry- weight of sediment on the idea of average unit weight of deposits. The full sediment yield for the amount of record lined by the survey can then be adequate the full weight of the sediment deposited within the reservoir and that that has passed out of the reservoir supported the lure potency during this approach reliable records could also be pronto and economically obtained on semipermanent basis. The density of deposited sediment varies with the composition of the deposits, location of the deposit among the reservoir, the activity characteristics of clay content and water, the age of deposit, etc. For coarse material (0.0625 millimetre and above) variation of density with location and age could also be unimportant.

Normally a time and house average density of deposited materials applicable for the amount below study is needed for locating the volume of deposits. For this purpose the unfree sediment for the amount below study would need to be classified in numerous fractions. Most of the sediment escape front obtaining deposited into the reservoir ought to be front the silt and clay fractions. In some special cases native estimates of densities at points within the reservoir could also be needed rather than average density over the total reservoir. The lure potency chiefly depends upon the capacity-in-flow magnitude relation however could vary with location of retailers and reservoir procedure. Computation of reservoir lure potency could also be created exploitation lure potency curves, like those developed by Brune and by Churchill (see IS: 12182-1987).

Life of a Reservoir

The dead storage provided in reservoir capability is allowed for geological phenomenon really all the sediment load will not get into dead storage. It encroaches upon live storage additionally. The encroachment and its distribution rely upon several factors like reservoir operation, depression characteristics, capability flow magnitude relation, sediment content within the flow etc. The helpful lifetime of a reservoir is taken until its capability is reduced to concerning two hundredth of the designed capability the speed of geological phenomenon is higher within the initial stages and it decreases with years.

5.2 Dams and its Types

The type of dam influences the design flood and spillway. Earth and rockfill dams have to be provided with ample spillway capacity.

Types of Dams

Most of the dams can be grouped into one of the following two categories:

- Embankment dams.
- Concrete dams.

Embankment dams include earth-fill dams and rock-fill dams. Concrete dams include gravity dams, arch dams and buttress dams. Preliminary designs and estimates will usually be required for different types of dams before one can decide the suitability or otherwise of one type of dam in comparison to other types. The cost of construction is the most important factor to be considered while making the final selection of the type of dam. Besides, the characteristics of each type of dam, as related to the physical features of the site and its adaptation to the purposes of the dam, as well as safety, and other relevant limitations are also to be considered for selecting the best type of dam for a particular site. The following are the important factors which affect the choice of the type of dam:

- Topography.
- Geology and foundation conditions.
- Material available.
- Size and location of spillway.

Topography of the site dictates the first choice of the type of dam. A concrete dam would be the obvious choice for narrow stream flowing between high and rocky abutments (i.e., deep gorges). Broad valleys in plains would suggest an embankment dam with a separate spillway.

Geological and hydrogeological characteristics of the strata which are to carry the weight of the dam determine the foundation conditions. Any type of dam can be constructed on solid rock foundations. Well-compacted gravel foundations are suitable for concrete gravity dams of small height, earth-fill, and rock-fill dams. However, effective cutoffs are required to check the foundation seepage. Silt or fine sand foundations can support concrete dams of small height and earth-fill dams. Problems of settlement, piping, and the foundation seepage are associated with this type of foundation. Non-uniform foundations containing different types of strata will usually require special treatment before any type of dam is constructed on such foundations.

If the construction materials to be used in large quantity for the construction of the dam are available in sufficient quantity within a reasonable distance from the site, the cost of the dam will be considerably reduced due to saving on transportation. If suitable soils for the construction of an earth-fill dam are locally available in nearby borrow pits, choice of an earthfill dam would be the most economical. The availability of sand and gravel (for concrete) near the dam site would reduce the cost of a concrete dam.

Spillway is a major part of any dam and its size, type, and the natural restrictions in its location will affect the selection of the type of dam. Spillway requirements are decided by the runoff and stream flow characteristics. As such, spillway on dams across streams of large flood potential can become the dominant part of the dam and put the selection of the type of dam to a secondary position. For large spillways, it may be desirable to combine the spillway and dam into one structure. This is possible only in concrete dams. Embankment dams are based on more conservative design assumptions and, hence, spillway is generally not constructed as part of the embankment. On the other hand, excavated material from a separate spillway can be advantageously used for the construction of an embankment dam.

Purpose of dam and operating conditions: The ungated spillway must be provided, except in certain circumstances when gated spillway can be provided.

5.2.1 Selection of the Type of Dam and their Classifications

Dams can be classified in various ways depending upon the purpose of the classification.

Classification According to Material used for Dam Construction

The dams classified according to the material used for the construction are: Earthen 'dams, Solid masonry gravity dams, Hollow masonry gravity dams, Rocklin dams, Steel dams, Timber dams, and R.C.C. Arch dams. They have already been explained in a previous article.

Classification of Dams According to Use

- The Detention Dams: They detain the food-waters temporarily in order to retard flood runoff and thus minimize the bad effects of the sudden flood. Detention dams are sometimes constructed to trap the sediment. They are often called debris dams.
- Diversion Dams: These small dams are used to raise the river water level, in order to feed an off-taking canal and or some other conveyance system. They are very useful as irrigation development works. A diversion dam is basically known as a weir or a barrage.
- Storage Dams: They are constructed in order to store the water during the periods of surplus water supply, to be used later during periods of deficient supply. The stored water shall be used in different seasons and for the different uses. They can be further classified depending upon the specific use of this water, like navigation, recreation, water supply, fish, electricity, etc.

Classification According to Hydraulic Designs

- Overflow Dams: They are designed to pass the surplus or extra water over their crest. They are often known as Spillways. They must be made of materials which won't be eroded by such discharges.
- Rigid Dams and Non-rigid Dams: Rigid dams are those which are constructed with the help of rigid materials like masonry, concrete, steel, timber, etc., while non-rigid dams are constructed of earth and rock-fill.
- Non-overflow Dams: They are the dams which are not designed to be overtopped. This design gives a wider choice of materials including earthfill and rockfill dams. Many a times, the overflow dam and the non-overflow dam are combined together to form composite single structure.

5.2.2 Selection of Dam Site

The selection of a site for constructing a dam should be governed by the following factors:

- Suitable foundations must be available.
- For economy, the length of dam must be as small as possible, and for a given height, it must store the maximum volume of water. It, therefore, that the river valley at the dam site should be narrow but should open out upstream to provide a large basin for the reservoir. A general configuration of contours for a suitable site is shown in below figure:

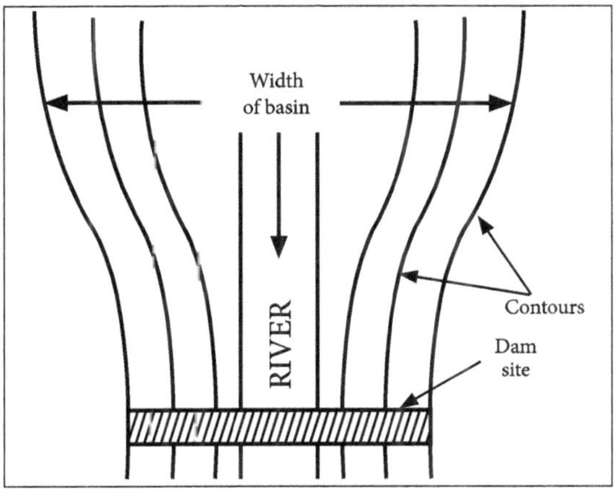

- The general bed level at dam site should preferably be higher than that of the river basin. This will reduce the height of the dam and will facilitate the drainage problem.
- A suitable site for the spillway should be available in the near vicinity. If the spillway is to be combined with the dam, the width of the gorge must be such as to accommodate both. The best dam site is one, in which a narrow deep gorge is separated from the flank by a hillock with its surface above the dam, as shown in below figure. If such a site is available, the spillway may be located separately in the flank, and the main valley spanned by an earthen or similar dam. Sometimes, the spillway and concrete masonry dam can be compositely spanned in the main gorge, while the flanks are in earth at low cost.
- Materials required for the construction need to be easily available, either locally or in the near vicinity, so that cost of transporting them is as low as possible.
- The reservoir basin must be reasonably water-tight. The stored water must not escape out through its side walls and bed.
- The value of land and property submerged by proposed dam should be as low as possible.

- The dam site should be easily accessible, so that it can be economically connected to important towns and cities by rails, roads, etc.
- The site for establishing labour colonies and a healthy environment need to be available in the near vicinity.

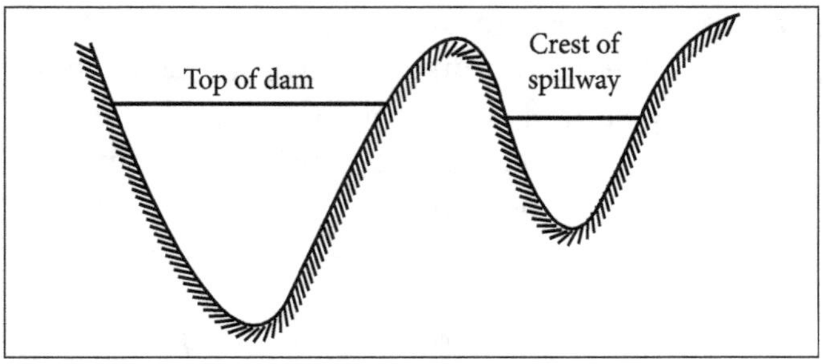

5.3 Gravity Dams

It is a masonry or concrete dam which resists the forces acting on it by its own weight. Its c/s is approximately triangular in shape:

- Straight gravity dam: A gravity dam that is straight in plan.
- Curved gravity plan: A gravity dam that is curved in plan.
- Curved gravity dam: It resists the forces acting on it by combining the gravity action and arch action.
- Solid gravity dam: Its body consists of a solid mass of the masonry or concrete.
- Hollow gravity dam: It has hollow spaces within its body.
- Most gravity dams are straight solid gravity dams.
- Concrete Gravity Dams: Weight holds the dam in place.

Lots of Concrete

These dams are heavy and massive wall-like structures of concrete in which whole weight acts vertically downwards.

As the entire load is transmitted on the small area of foundation, such dams are constructed where rocks are competent and stable.

Bhakra Dam is the highest Concrete Gravity dam in Asia and is the second highest in the world. Bhakra Dam is across river Sutlej in Himachal Pradesh.

Concrete gravity dam.

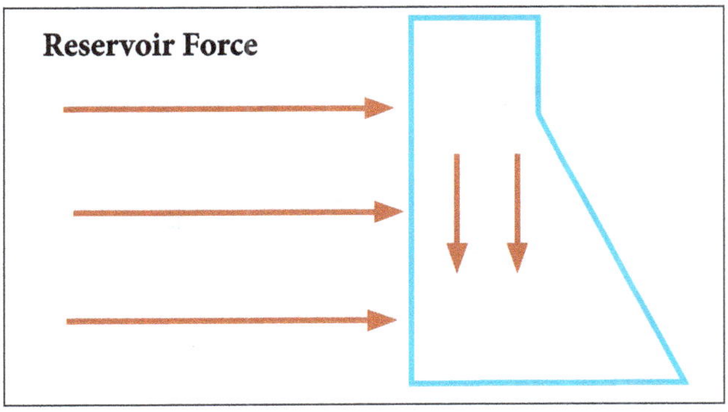

Reservoir force.

The construction of this project was started in the year 1948 and was completed in 1963. It is 740 ft above the deepest foundation as straight concrete dam being more than three times the height of Qutub Minar.

Length at the top 518.16m; width at the base 190.5m and at the top is 9.14m Bhakra.

Dam is the highest Concrete Gravity dam in Asia and the Second Highest in the world.

A gravity dam is any solid structure made of concrete or masonry built across a river to create a reservoir on its upstream. The section of gravity dam is approximately triangular in shape, with its apex at the top and the maximum width at the bottom. The section is so proportioned that it resists the various forces acting on it by its own weight are usually consist of two sections; such as, the non-overflow section and the overflow section or spillway section are particularly suited across gorges with very steep side slopes and the earth dams might slip and are usually cheaper than the earth dams if suitable soils are not available for their construction where good foundations are available, the gravity dams shall be built up to any height. It is the most permanent one, and it requires only little maintenance. The most ancient gravity, dam on the record was built in Egypt more than 400 years B.C. of uncemented masonry.

Bhakra Dam.

Gravity dam.

Basic Definitions

- Length of the dam: It is the distance from one abutment to other, measured along the axis of the dam at the level of top of the dam.
- Axis of the dam: It is the line of o upstream edge of the top of the dam. The axis of dam in plan is called the base line of dam. And they are usually straight.
- Toe and Heel: The toe of the dam is the downstream edge of the base, and the heel is the upstream edge of the base.
- Structural height of the dam: It is the difference in elevations of the top of the dam and the lowest point in the excavated foundation. It does not include the depth of special geological features of foundations like narrow fault zones below the foundation. In general, the height of the dam means its structural height.
- Hydraulic height of the dam: It is equal to the difference in elevations of highest controlled water surface on the upstream of the dam and the lowest point in the river bed.
- Maximum base width of the dam: It is the maximum horizontal distance between the heel and toe of the maximum section of the dam in the middle of valley.

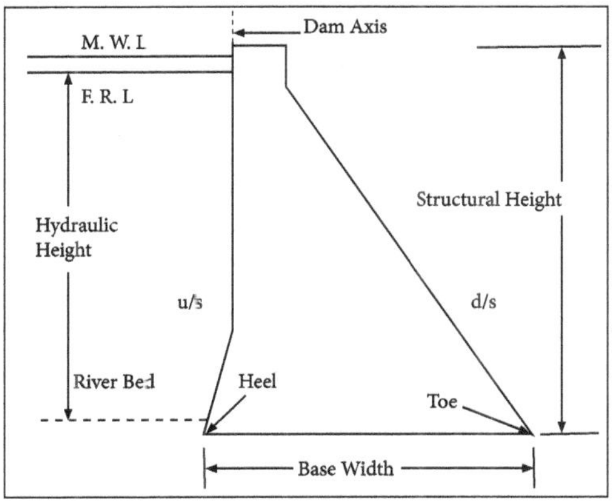

5.3.1 Forces Acting on a Gravity Dam

- Water pressure.
- Weight of the dam.
- Wave pressure.
- Uplift pressure.
- Earthquake forces.
- Earth and Silt pressure.
- Wind pressure.
- Ice pressure.
- Thermal loads.

These forces fall into the two categories as:

- Forces like weight of the dam and the water pressure, which are directly calculable from the unit weights of the materials and the properties of fluid pressures.
- Forces like uplift, earthquake loads, silt pressure and ice pressure, which can only be assumed on the basis of assumption of varying degree of reliability.

It is in the estimation of the second category of forces that special care need to be taken and reliance placed on available data, experience and the judgment.

It is convenient to compute all forces per unit length of the dam.

Weight of Dam

- Main stabilizing force in the gravity dam

- Dead load = weight of concrete or masonry or both + weight of such appurtenances as piers, gates and bridges.
- Weight of the dam per unit length is equal to the product of the area of cross-section of the dam and specific weight of the material.
- Unit weight of concrete and masonry varies considerably depending upon the several materials that go to make them.

For convenience, the cross-section of the dam is divided into any simple geometrical shapes, like rectangles and triangles, for the computation of weights. The areas and centroids of those shapes may be easily obtained. Hence the weight components W_1, W_2, W_3 etc. may be found along with their lines of action. The total weight W of the dam acts at C.G. of the section:

Water Pressure

Water pressure on the upstream face is the main destabilizing force acting on the gravity dam. Tail water pressure helps in the stability.

Although the weight of water changes slightly with temp., the variation is usually ignored. Unit Mass of the water is taken as 1000 kg/m³ and the specific weight = 10 kN/m³ instead of 9.81 kN/m³.The water pressure acts normal to the face of dam. It is convenient to determine the components of the forces in the horizontal and vertical directions instead of the total force on the inclined surface directly.

The water pressure intensity p (kN/m²) varies linearly with depth of water measured below the free surface y (m) and is expressed as follows,

$$P = \gamma_w y.$$

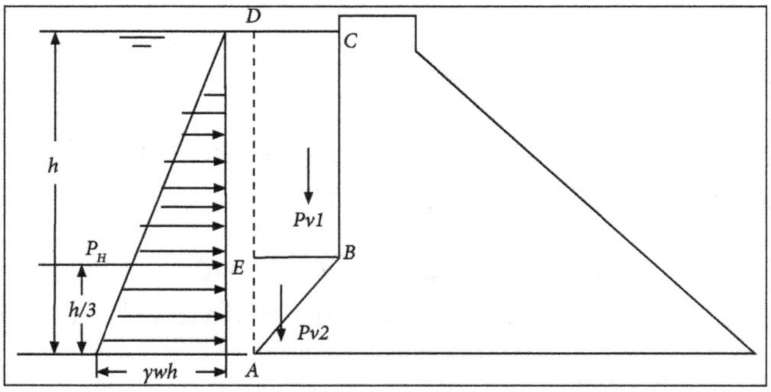

Water Pressure U/s Face Vertical

When the upstream face of the dam is vertical, the water pressure diagram is triangular in shape with a pressure intensity of $\gamma_w h$ at the base, where h is the depth of the water. The total water pressure per unit length is horizontal and is given by,

It acts horizontally at the height of h/3 above the base of the dam.

U/s Face Inclined

When the upstream face ABC is inclined or partly vertical and partly inclined, the force due to the water pressure may be calculated in terms of the horizontal component PH and the vertical component PV. The horizontal component is given as the earlier and acts horizontal at a height of (h/3) above the base. The vertical component PV of the water pressure per unit length is equal to the weight of the water in the prism ABCD per unit length. For convenience, the weight of the water is found in two parts P_{V1} and P_{V2} by dividing a trapezium ABCD into rectangle BCDE and triangle ABE.

Hence the vertical component $PV = PV_1 + PV_2$ = weight of the water in BCDE + weight of the water in ABE. The lines of action of PV_1 and PV_2 will pass through the respective centroids of the rectangle and triangle.

Uplift Pressure

- Water has the tendency to seep through pores and fissures of the material in the body of the dam and the foundation material, and through the joints between the body of the dam and its foundation at the base. The seeping water exerts pressure.
- A portion of the weight of dam may be supported on the upward pressure of water; thus net foundation reaction due to vertical force will reduce.
- The uplift pressure is defined as the upward pressure of water as it flows or seeps through the body of dam or its foundation.
- One school of thought recommends that a value one-third to two-thirds of the area must be considered as effective over which the uplift acts.
- The area over which the uplift pressure acts has been a question of investigation from the early part of the century.

Code of Indian Standards (IS: 6512-1984)

- There are two constituent elements in uplift pressure: the area factor or the percentage of area on which uplift acts and the intensity factor or the ratio which the actual intensity of uplift pressure bears to the intensity gradient extending from head water to tail water at various points.
- The total area should be considered as effective to account for uplift.
- The pressure gradient shall then be extending linearly to heads corresponding to reservoir level and tail water level.

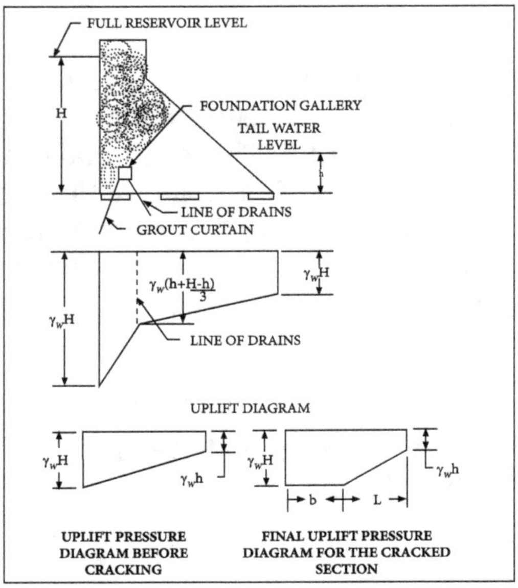

In case of drain holes: the uplift pressure in the line of drains exceeds the tail water pressure by one-third the differential between the reservoir and tail water heads. The pressure gradient shall then be extended linearly to heads corresponding to reservoir level and tail water level.

In case of a crack: The uplift is assumed to be reservoir pressure from the u/s face to the end of the crack and from there to vary linearly to the tail water or drain pressure.

In absence of the line of drains and for the extreme loading conditions F and G, the uplift can be taken as varying linearly from the appropriate reservoir water pressure at the u/s face to the appropriate tail water pressure at the d/s face.

Uplift pressures are not affected by earthquakes.

Earth and Silt Pressure

Gravity dams area unit subjected to earth pressures on the downstream and upstream faces wherever the muse trench is to be backfilled. Except within the abutment sections in specific cases, earth pressures have sometimes a minor impact on the steadiness of the structure and will be unnoticed. Silt is treated as a saturated cohesion less soil having full uplift and whose price of internal friction isn't materially modified on account of submerging.IS code recommends that a) Horizontal silt and water pressure is assumed to be cherish that of a fluid with a mass of 1360 kg/m^3, and b) Vertical silt and water pressure is set as if silt and water along have a density of 1925kg/m^3.

Ice Pressure

Ice expands and contracts with changes in temperature. In a reservoir utterly frozen over, a call the air temperature or within the level of the reservoir water might cause the gap from cracks that afterward fill with water and freezed solid once succeeding rise in temperature happens, the ice expands and, if restrained, it exerts pressure on the dam.

Good analytical procedures exist for computing ice pressures, however the accuracy of results relies upon sure physical information that haven't been adequately determined.

Ice pressure is also provided for at the speed of 250 kPa applied to the face of dam over the anticipated space of contact of ice with the face of dam.

The problem of ice pressure within the style of dam isn't encountered in Bharat except, perhaps, during a few localities.

Wind Pressure

Wind pressure will exist however is rarely a major consider the planning of a dam. Wind masses might, therefore, be ignored.

Thermal Load

The cyclic variation of air temperature and also the radiation on the downstream facet and also the reservoir temperature on the upstream facet have an effect on the stresses within the dam.

Even the deflection of the dam is most within the morning and it goes on reducing to a minimum price within the evening.

Measures for temperature management of concrete in solid gravity dams area unit adopted throughout construction.

Thermal aren't important in gravity dams and will be unnoticed.

5.3.2 Causes of Failure Of a Gravity Dam

A gravity dam may fail on account of overturning. For a gravity dam to be safe against overturning, the dimensions of the dam should be such that the resultant of all the forces intersects the base of the dam within its middle-third portion. Consider any horizontal section (including the base) of a gravity dam and the resultant of all the forces acting on the dam above the section. If the line of action of this resultant passes outside the downstream edge of the section, the dam would overturn. However, the section of a gravity dam is such that the line of action of the resultant force is within the upstream and downstream edges of the section and overturning would never results. But, if the line of action of the resultant passes sufficiently outside the middle-third of the horizontal section, it may cause crushing of the downstream edge of the section. This would reduce the effective width and, hence, the sliding resistance of the section and may cause the resultant to pass outside the dam section. Further, when the resultant passes downstream of the middle-third of the horizontal section, it induces tensile stresses at the upstream edge of the section. These tensile stresses may cause cracks in the dam section which would result in increased uplift pressure. The stabilising forces would, thus, be reduced. It follows, therefore, that before a gravity dam overturns bodily, other types of failures, such as crushing of toe material, sliding, cracking of the material due to tension, and increase in uplift may occur. A gravity dam is considered safe against overturning if the criteria of: (i) no tension on the upstream face, (ii) adequate resistance against sliding, and (iii) suitable quality and sufficient strength of concrete/masonry of dam and its foundation are satisfied.

Concrete and masonry are relatively weak in tension and as such the design of a gravity dam should ensure that there are no tensile stresses anywhere in the dam section. In very high gravity dams, however, if it becomes difficult to ensure such a condition, one may allow mall tensile stresses not exceeding 50 N/cm^2 under the most adverse condition of loading.

The horizontal forces acting on a dam above any horizontal plane may cause failure of the dam due to sliding if these driving forces are more than the resistance to sliding on the plane. The resistance to sliding is due to the frictional resistance and shearing strength of the material along the plane under consideration. The shear-friction factor of safety, Fs, which is a measure of stability against sliding or shearing, can be expressed as follows:

Where,

C = unit cohesion,

$$F_8 = \frac{CA + \sum W\mu}{\sum H}$$

A = area of the plane considered (A can be replaced by the width of the plane, if one considers unit length of the dam).

$\sum W$ = sum of all vertical forces acting on the plane.

μ = coefficient of internal friction.

$\sum H$ = sum of driving shear forces i.e., resultant horizontal forces.

The shear-friction factor of safety can be used to determine the stability against sliding or shearing at any horizontal section within a dam, its contact with the foundation or through the foundation along any plane of weakness. The minimum allowable values of F_s for gravity dam are 3.0, 2.0, and 1.0 for the usual, unusual, and extreme loading combinations, respectively. The value of F_s for any plane of weakness within the foundation should not be less than 4.0, 2.7, and 1.3 for the usual, unusual, and extreme loading combinations, respectively.

Generally, the acceptable factor of safety against overturning and shear for normal or usual loading condition is taken as 2.0. The corresponding value for extreme loading condition is 1.25. The acceptable value of the sliding factor (= the ratio of the sum of the horizontal forces and the sum of the vertical forces) is the ratio of the coefficient of static friction and the chosen factor of safety. The maximum allowable compressive stress for concrete in a gravity dam should be less than the specified compressive strength of the concrete divided by 3.0, 2.0, and 1.0 for usual, unusual, and extreme load combinations, respectively. The compressive stress should not exceed1035 N/cm2 and 1550 N/cm2 for usual and unusual load combinations, respectively .The maximum allowable compressive stress in the foundation should be less than the compressive strength of the foundation divided by 4.0, 2.7, and 1.3 for usual, unusual, and extreme load combinations, respectively. These values of factor of safety are higher than those for concrete so as to provide for uncertainties in estimating the foundation properties.

5.3.3 Elementary or Theoretical Profile and Practical Profile of Gravity Dam

The elementary profile of a gravity dam is a triangular section having zero width at the water level where hydrostatic pressure is zero and maximum base width B, where maximum hydrostatic pressure head acts.

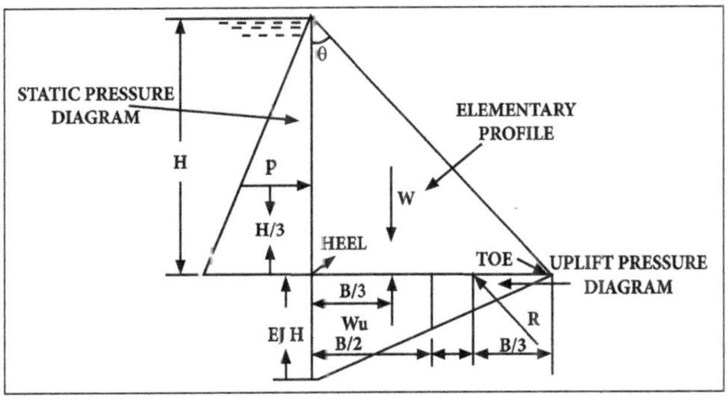

The elementary profile is not possible from practical point of view as dam requires:

- Road at the top.
- Additional Load due to Roadway Top.
- Free Board.

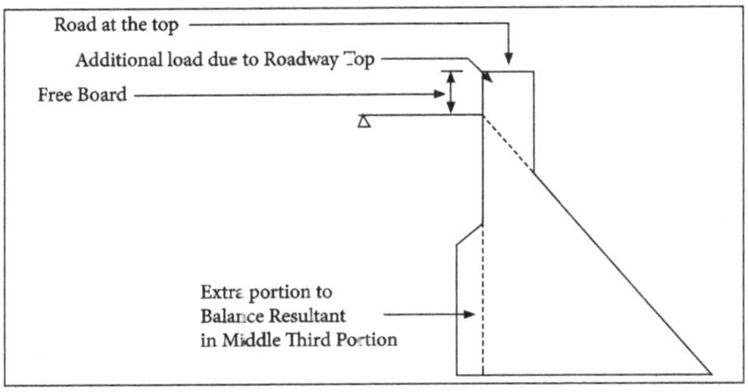

Practical Profile of Gravity Dam

- Hence the practical profile will be different from the theoretical profile.
- Due to the provision of the above three, the resultant will act outside the outer middle third and tension at the heel and compression at the toe will occur. To eliminate these tension and compression, some extra concrete will be essential in the upstream face.

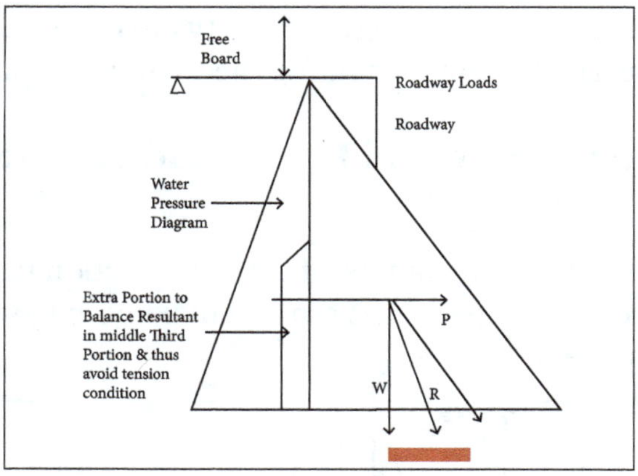

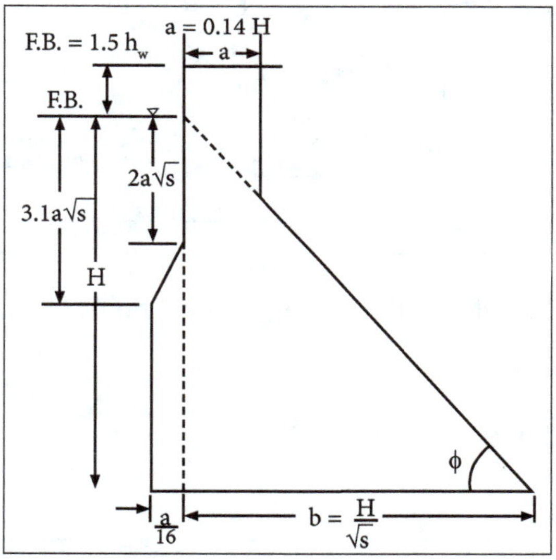

5.3.4 Limiting Height of a Gravity Dam and Stability Analysis

The only variable in the expression for the principal stress σ1 at the toe is H. The maximum value of this principal stress should not exceed the allowable stress σ per for the material i.e. σ 1≤ σ per. In the limiting case,

$$\sigma_1 = H\gamma_w (S_c - C + 1) = \sigma_{per}$$

From which, the limiting height H_{lim} is given by,

$$H_{lim} = \sigma_{per} / \gamma_w (S_c - C + 1)$$

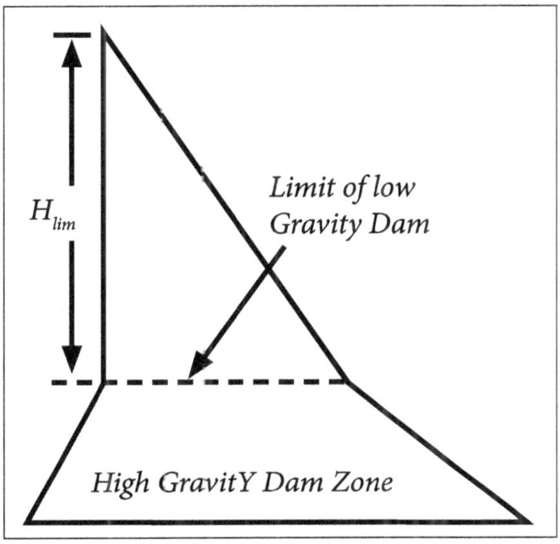

For finding the limiting height $H_{lim,}$ it is usual not to consider the uplift.

Hence putting,

$$H_{lim} = \sigma_{per} / \gamma_w (S_c + 1)$$

If the height of the dam is more than $H_{lim,}$ the maximum compressive stress will exceed permissible stress and that condition is undesirable. This equation for the limiting height defines the distinction between a low and a high gravity dam. A low gravity dam is the one in which the height H is less than H_{lim} so that maximum compressive stress is not greater than the allowable stress. For a concrete dam (S c = 2.40 and σ per = 3.0 N/mm²), the limiting height is about 88 m. If higher grade concrete (σ per ≥= 3.0 N/mm2) is used then the limiting height would be more. When action will have to be given extra s the height of the dam to be constructed is more than that $H_{lim,}$ the dam is known as high gravity dam. For such type of dam, the section will have to be given extra slopes to the upstream and downstream sides, below the limiting height, to bring the compressive stress within the permissible limits, as illustrated in Figure above.

Steps in the Stability Analysis

1. Determination of Forces and their Moments

All the force acting on the dam should be written first. For earthquake, the direction of acceleration should be mentioned, so as to determine the force correctly.

All the forces are calculated as per criteria and formulae detailed earlier. The moments are taken with respect to upstream heel point, for facility of calculations. In the example for which data is given the different loads have been calculated as below:

i. Dead weights: The weight of the dam section can be found by dividing the section into simpler geometrical shapes, like triangle, rectangle, trapezium etc. The area of each part can be easily calculated and multiplying by density of concrete as 2.4 t /m^3 the weight of each part can be calculated. This weight acts vertically downwards through the center of gravity which can be found easily for each geometrical figure. Thus the weight W1 in the dam section pertains to the top triangle of width 6.1 m and height 6.72 m (width of dam section normal to the plane of figure is taken as 1.0 metre). The weight is given by $\frac{2.4\times6.1\times6.72}{2}$ i.e 49 tons acting on a distance 2 /3 X 6.1 = 4.06m from upstream heel is therefore 49x 4.16 = 198 tm.

We consider all clockwise moments about heel. The moment of this weight W1 about heel as negative and all anticlockwise moments as positive. Thus the moment due to all dead weights are negative.

Water Thrust: The water pressure acts horizontally on the upstream face and is given by ½w. h2 In this case, water thrust = ½ x 52 x 52x 1 = 1350 t . It acts at a distance of h /3 i.e 52 /3 m from heel. Hence moment is negative and equals to 1350 x 52 /3 = - 23400 t.m.

Uplift Forces: The uplift intensity at the upstream heel end is which i.e. 52 t/sq. m and on the downstream it will be whd if hd is the tail water depth. In this case, tail water is taken as zero and hence uplift varies linearly from 52 t / sq m to 0 value at toe (when drains are choked). If drains are working then the balance uplift at the position of drain is 52/3 i.e 17.33t. The uplift pressure diagram can be conveniently divided into triangles and rectangles and the uplift pressure on the whole base width may be calculated similar to dead loads. The uplift on any portion will act from the centre of gravity of that part. Obviously these force are acting opposite to dead weight and are therefore, taken as negative. The moments about heel are anticlockwise and are therefore taken as negative. The moments about heel anticlockwise and hence positive.

Hydrodynamic Pressure due to Earthquake: The horizontal earthquake will produce a force given by,

Pe = 0.726 Pe. h., where,

Pe = C a w h

$$= \left(0.73\frac{90^\circ}{90\circ}\right)0.20\times1\times52(\text{for base})$$

(Since = 0.20 and upstream face is vertical)

P = 0.726 x 0.73 x 0.20 x 52 x 52 = 286 t

The moment about heel is given by 0.3 peh

= – 0.3 x 0.73 x 0.20 x 523= – 6150 t.m (for base)

It the earthquake acceleration is acting upstream the force is acting horizontally downstream and moment is negative and vice versa.

Inertial Forces due to Earthquake

Effect of Vertical Earthquake: This will effect the inertial forces of the concrete mass and water. The forces will increase, if the earthquake acceleration is acting vertically up and decrease if the acceleration is acting downward. Thus the dead weighs and water pressure will be modified accordingly by multiplying forces and moments due to masonry and water by (1± v) where v is the vertical earthquake coefficient.

Effect of Horizontal Earthquake: It will create acceleration in the dam body and consequently an inertial force which is dead weight of masonry multiplied by the earthquake coefficient, is created. This force acts horizontally upstream or downstream depending on the acceleration direction of the earthquake), through the centre of gravity of the dam section. Its moment can be taken about the heel.

Silt Load: The effect of silt of horizontal pressure is to consider the density of water as 1.36 t/m3instead of 1 t/m^3 to 0.36 t/m3from heel i.e. 52/3 m from base giving a moment of -8450 t.m.

Since upstream face of dam is vertical, no vertical silt load has been considered.

Wave Pressure: The total pressure due to waves is given by, 2.0 w hw2case, P = 2.0 x 1.5^2 calculated from Molitor's formulae).

Since the pressure due to waves acts at height of (3 h_w/8) = 3 x1. 5/8 = 0.6 m from reservoir surface, lever arm of the pressure about upstream heel = 52 + 0.6 = 52.6 m. The moment about heel is negative and is equal to 4.5 x 52.6 = 236 t.m.

2. Tabulation of Forces and Moments

Having determined forces and moments, they can be tabulated and summed up. Dividing total moments (algebraic addition of positive and negative moments) by the total vertical force, the point of application of the resultant on the dam bottom line from heel is determined. Subtracting half bottom width from the above, the eccentricity of the resultant force is found.

3. Determination of Stresses

The stresses, z and principal stresses p on the upstream and downstream faces can be calculated from the formulae. The calculations for each case for the problem are given at the end of each table.

5.3.5 Galleries in Gravity Dams and Grouting

General

Galleries are provided in a dam for different purposes viz. grouting and drainage, operation of gates and control equipment, housing hoisting pumping and other equipment for measurement of structural behavior of the dam. In addition there can be other types of openings like sluices, conduits etc. Which run full of water. Thus we can divide galleries (i.e. A formed opening in a dam) as (i) hydraulic like sluices, outlets, penstocks etc. and (ii) galleries which can be termed as control and inspection passages.

Galleries can be further subdivided according to their alignment viz.

1. Parallel to the Axis of Dam

In this category, we can include:

Foundation and Drainage Gallery: This gallery generally extends the length of the dam near the rock surface, conforming in elevation to the transverse profile of the canyon; in plan it is near to the axis of dam. It is from this gallery that the holes for the main grout curtain are drilled and grouted and also from which foundation drain holes are drilled. Its size is approximately 1.5 x 2.2 or 1.8 x 2.4 m.

Drainage Gallery: A supplementary drainage is sometimes located further downstream, about 2 / 3rd of the base width from the upstream face, for the purpose of drilling and draining the downstream portion of the foundation. This gallery usually extends only through the deepest portion of the dam.

Gate Gallery: Location and size depends on the need and size of the equipment to be housed. It usually connects hoisting chambers just over the control gates. Examples may be sluice-operating gallery, penstock-operating gallery etc.

Grouting Galleries: If it is difficult to grout contraction joints from the face of the dam the grout piping system should be arranged so as to locate the supply, return and vent headers in galleries placed near the top of each 15m life of concrete. The piping system for artificial cooling of the blocks should also be arranged to terminate in these galleries.

Inspection Galleries: Even some galleries are provided for the inspection purposes.

Normal to Dam Axis: In this group we can include adits, which are approaches to the different galleries. These admits also sometimes locate measuring boards of the different equipment's pump and so on.

Vertical and Inclined shafts, These include:

- Shafts for housing elevators, plumb bob etc.

- Inclined shaft to connect two galleries or one gallery at two elevations.

Grouting the Foundation

The foundation grouting may be divided into:

- Consolidation Grouting.

- Curtain Grouting.

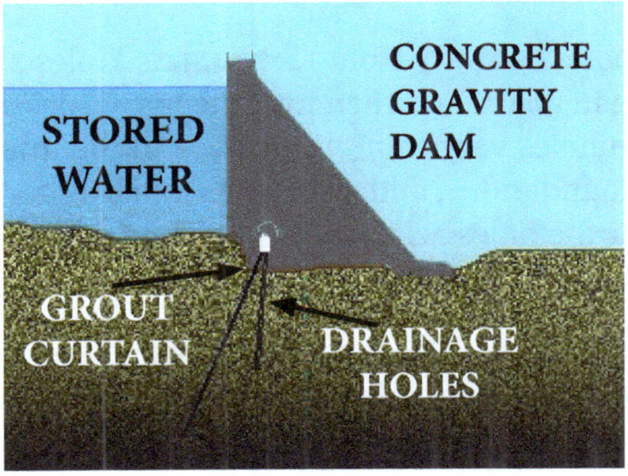

Consolidation Grouting

- The entire foundation of the dam is consolidated by means of grouting. For this purpose, shallow holes are drilled through foundation rock. The depths of these holes usually vary between 10 to 15 m. They are situated at about 5 to 20 m apart, in general area of the heel of the dam. After the drilling of holes, mixture of cement and water known as grout is forced into the holes at low pressure for about 30 to 40 N/ cm^2.
- This is accomplished, even before any concreting for the dam section is been laid. This low pressure grouting will surely result in a general consolidation of the foundations. These low pressure grout will later serve the purpose of a cut-off against leakage of high pressure grout which is to be used after some concreting of the dam has taken place.

Consolidation Grouting.

Curtain Grouting

- It helps in forming the principal barrier or a curtain against seepage through the foundations, and reduces the uplift pressure. To accomplish high pressure grouting, and relatively deeper holes are drilled near the heal of the dam.
- The spacing of holes can vary from 1.2 to 1.5 m. Holes are , drilled and grouted at about 10 to 12 m apart, and then intermediate holes are drilled and grouted. The depth of the holes vary from 30 to 40 % of the total upstream water head for strong rock foundations, and shall be as much as 70 % of the water head for poor rocks. After the holes are drilled, a mixture of cement and water is forced into holes under high pressure.
- The grouting pressure shall be kept as high as possible without lifting foundation strata. Generally, the foundation pressure used in this high pressure grouting is equal to 2.5 D N/ cm 2, where D is the depth of grouting in metres below the surface. This grouting is usually done in stages of depth equal to 15 m or so, and carried out only after some portion of the dam section has been laid.
- This grouting has to be accomplished from the foundation gallery or from other galleries within the dam. It may also be done from the upstream face of the dam, if possible, In certain special cases, this grouting may have to be accomplished from tunnels driven into the foundation rock below the dam.

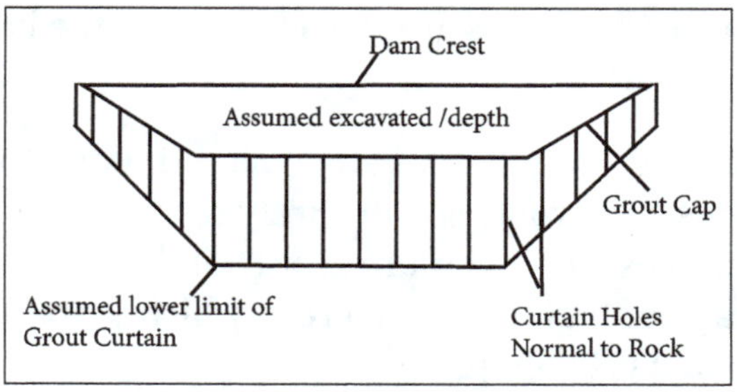

Curtain Grouting.

Chapter 6
Earth Dams and Spillways

6.1 Earth Dams and its Types

Earth dams for the storage of water for the purpose of irrigation have been built since the earliest times. These dams were limited in height but not necessarily in extent. Now Earth dams are being built to unprecedented heights. Sites which have been considered unfit earlier for the construction of darns are now being exploited. The Development of soil mechanics, i.e., the study of behavior of earth dams, and the development of better construction techniques have been helpful in creating confidence to build higher dams with improved designs and more details. The result is that, today the highest dam in the world is an earth dam. The highest earth/rockfill dams in the world are Roguni U.S.S.R (335 m) Nurek, U.S.S.R. (300 m), Mica, Tehri India (260 m), Canada (244 m) and Oroville, U.S.A. (235 m).

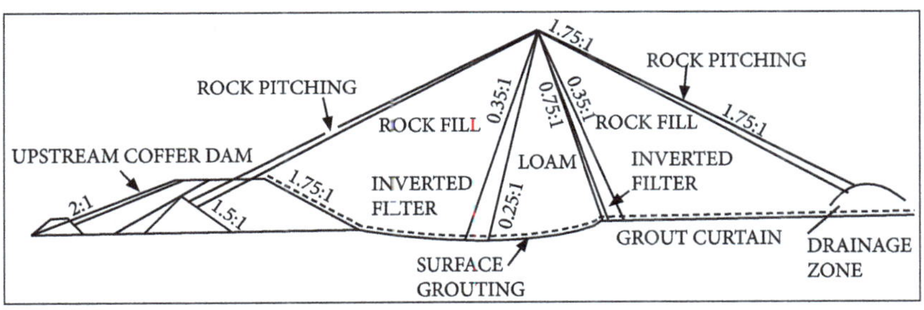

Earth dam.

In spite of these developments it is very difficult to establish mathematical solutions to the problems of design and components are still guided by experience and judgment. For a realistic design of an earth dam it is important that the foundation conditions and materials of construction are thoroughly investigated. It is also necessary that the controlled methods of construction are used to achieve necessary degree of compaction at the predetermined moisture. The discussion in the text is limited to the design procedure for earth dams which are rolled fill type of construction. This type of construction is being used almost entirely for the construction of earth dams to the exclusion of the hydraulic and the semi-hydraulic fills. In this type, the major portion of the embankment is built in successive mechanically compacted layers of 150 mm to 220 mm thickness each.

Classification of Earthfill Dams

Earthfill dams are classified by number of factors.

Based on the method of construction:

- Rolled fill earth dams.
- Hydraulic fill dam.

Based on mechanical characteristics of earth materials making the section of the Dam:

- Homogeneous earth dams.
- Non-Homogeneous earth dams:
    - Non-homogeneous with inclined and impervious zone of artificial material.
    - With impervious zone of soil and also with low permeability.
    - With central core soil material of low permeability.
    - With a central thin diaphragm of impervious material.

Rolled Fill Earth Dams

In this type of dams, the successive layers of the moistened or damp soils are laid one over the other. Each layer that are not exceeding 20 cm in thickness is properly consolidated at optimum moisture content, only then the next layer laid.

Hydraulic Fill Dams

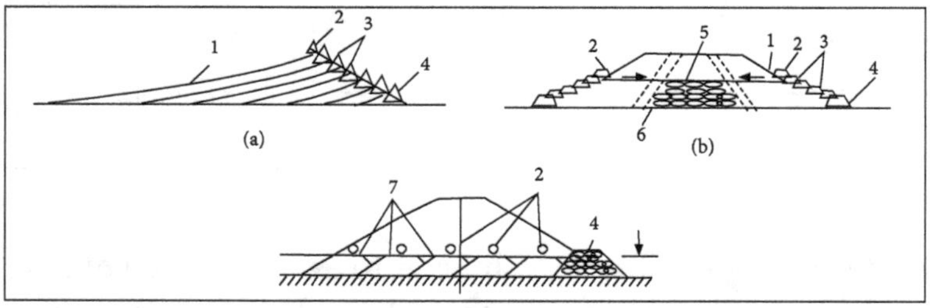

In this type of dams, the construction, the excavation, the transportation of the earth is done by the hydraulic methods. Outer edges of the embankments are kept slightly higher than the middle portion of the each layer. During construction, a mixture of the excavated materials in the slurry condition is pumped and discharged at the edges. This slurry of the excavated materials and water consists of coarse and fine materials. When it is discharged near the outer edges, the coarser materials settle first at the edges, while the finer materials move to the middle and then settle there. Fine particles are deposited in the central portion to form a water tight central core. In this method, compaction is not required.

Homogeneous Earthen Dam

These dams are constructed with the uniform and homogeneous materials. It is suitable for low height dams. These dams are generally constructed with soil and grit mixed in the proper ratios. The seepage action of such dams are unfavourable, therefore for safety in case of rapid drawdown, the upstream slope is kept relatively flat.

Homogeneous section is modified by constructing the rock toe at the downstream lower end and providing horizontal filter drain.

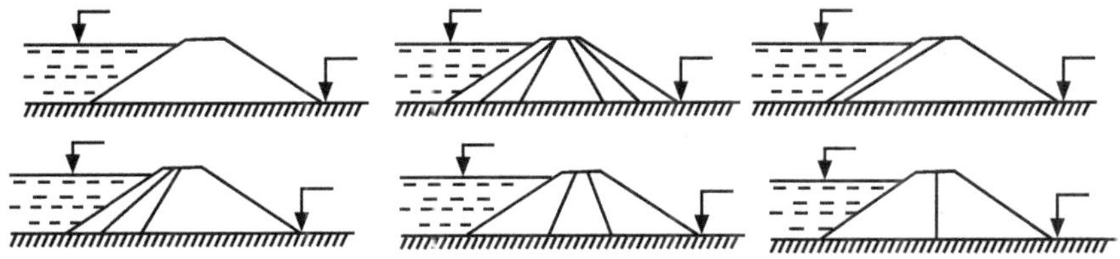

Zoned Earth Dams

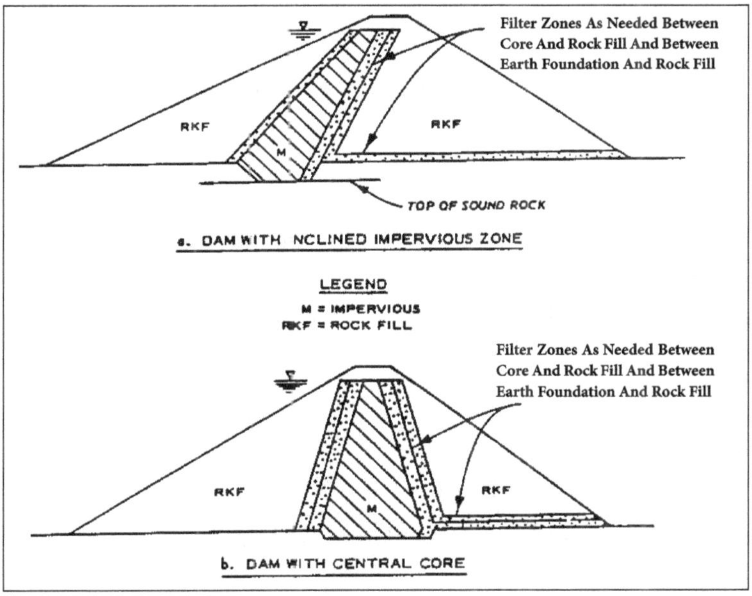

These are the dams with the central portions termed as core or hearting made from the materials which are relatively impervious. The thickness of the core wall is made sufficiently thick to prevent it from leakage of water through the body of dam.

Dam with a Diaphragm

This type of dam is constructed with the pervious materials, with a thin impervious diaphragm in the central part to prevent it from seepage of water. The thin impervious diaphragm can be made by means of impervious clayey soil, cement concrete or masonry or any of the impervious material. The diaphragm may be constructed in the central portion or on the upstream face of the dam. The main difference in zoned and diaphragm type of dams depend on the thickness of the impervious core or diaphragm. The thickness of the diaphragm is not more than 10 m.

Design Criteria of the Earth Dams

Based on experience of the failures, following main design criteria can be laid down for the safety of an earth dam.

1. To prevent hydraulic failures the dam should be so designed that erosion of the embankment is always prevented.

This implies that the following conditions are strictly satisfied:

- Spillway capacity is sufficient to pass a peak flow.
- Overtopping by wave action at maximum water level is prevented.
- The original height of structure is sufficient to maintain the minimum safe freeboard after settlement has occurred.
- Erosion of embankment due to the wave action and the surface run-off does not occur.
- The crest should be wide enough to withstand wave action and earthquake shock.

2. To prevent the seepage failures, the flow of water through the body of the dam and its foundation must not be sufficiently large in quantity to defeat the purpose of the structure nor at a pressure sufficiently high to cause piping.

This implies that:

- Quantity of seepage water through the dam section and foundation should be limited.
- The seepage line should be well within the downstream face of the dam to prevent sloughing.
- Seepage water through the dam or foundation should not remove any particle or in other words cause piping. The driving force depends upon the pressure gradient while the resisting force depends upon the strength characteristics of the boundary material.
- There should not be any leakage of water from the upstream to downstream face. Such leakage may occur through conduits, at joints between earth and concrete sections or through the holes made by aquatic animals.

3. To prevent the structural failures, the embankment and its foundation should be stable under all the conditions.

This implies that:

- The upstream and downstream slopes of embankment must be stable under all loading conditions to which they shall be subjected including the earthquake.
- The foundation shear stresses must be within the permissible limits of shear strength of the material.

6.1.1 Causes of Failures of Earth Dams

Just Like most other damages to engineering structures, earth dam failures are mostly caused due to improper design frequently based on insufficient investigations, and the lack of care in the construction and the maintenance.

Failures of earth darns can be grouped into the following basic causes:

- Hydraulic failures.
- Seepage failures.
- Structural failures.

Hydraulic Failures

They are responsible for about one third of the failure of the dams and are produced by the surface erosion of dam by the water. These includes wash-outs from overtopping wave erosion of the upstream face, scour from the discharge of the spillway etc. and erosion from the rainfall.

Seepage Failures

Seepage of water through the foundation or embankment are accountable for more than one third of the earth dam failures. Seepage is inevitable in all earth dams and no harm is done. Uncontrolled seepage; can cause erosion within the embankment or in the foundation which can lead to piping. Piping is progressive erosion which develops through under the dam. It begins at the point of concentrated seepage where the gradients are essentially high to produce the erosive velocities. If forces resisting erosion are less than those' which tend to cause, the soil particles are then washed away and that causes piping failure.

Seepage failures are generally caused by:

- Pervious foundations.
- Leakage through embankments.
- Conduit leakage.

- Sloughing.

Pervious Foundations: Presence of strata and lenses of sand or gravel of high permeability or cavities and fissures in the foundation can allow concentrated flow of the water from the reservoir resulting in piping. The Presence of buried channels under seat of dam are also responsible for this.

Leakage through Embankments

The following are common causes of embankment leaks which leads to piping:

- Poor construction control which includes insufficient compaction adjacent to outlet conduits and the poor bond between the embankment and the foundation or between the successive layers of embankment.
- Cracking in the embankment or in the conduits caused by the foundation settlement.
- Animal burrows.
- Shrinkage and dry cracks.
- The Presence of roots, pockets of gravel or boulders in the embankment.

Conduit Leakage: Conduits through the dam are responsible for nearly one third of seepage failure and more than one eighth of all the failures. Failures are of two types 1) contact seepage along the outside of the conduit which develops into piping and 2) seepage through leaks in the conduit which may also develop into piping. Contact seepage along the conduit wall is caused either by a zone of poorly compacted soil or the small gap between the conduit and remainder of the embankment. Seepage through poorly compacted zones soon develops causing piping. Conduit cracking is caused by the differential settlement or by overloading from embankment.

Sloughing: The Failure due to the sloughing takes place where the downstream portion of the dam becomes saturated either due to the choking of filter toe drain, or due to the presence of highly pervious layer in the body of dam. The process starts when a small amount of the material in the downstream toe is eroded and produces a small slide. It leaves a relatively steep face which becomes saturated by seepage from the reservoir and slumps again, forming a higher and more unstable face. This process is continued until the remaining portion of the dam is too thin to hold the water pressure and then complete failure occurs.

Structural Failures

Structural failures of the embankment or its foundation account for about one fifth of the total number of failures. Structural failures will result in slides in foundation or embankment due to the various causes as explained below.

Foundation Failures

Faults and seams of weathered rocks, soft clay, shale, strata are responsible for the foundation failure in which top of the embankment cracks and subsides and the lower slope moves outward and large mud waves are formed beyond the toe. Another form of foundation failure happens because of the excessive pore water pressure in confined seams of silt or the sand. Pore water pressure in the confined cohesion less seams, artesian pressure in the abutments or consolidation of the clays interbedded with the sands or silt, reduces the strength of the soil accordingly to the extent that it cannot be able to resist the shear stresses induced by embankment. The movement develops very rapidly even without warning. Excess settlement of foundation may cause cracking of the embankment.

Types of Failures: Earth Dams

Slides in Embankment

An embankment is subjected to shear stresses imposed by the pool fluctuations, seepage or earthquake forces. Embankment slides might occur when the slopes are too steep for shear strength of the embankment material to resist the stresses imposed. Generally the movement develops slowly and is preceded by cracks on the top or the slope near the top. Failure of this type are usually due to the faulty design and the construction. In high dams slope failure can occur during dissipation of pore pressure just after construction. The upstream slope failure occurs due to sudden drawdown as shown in figure. The downstream slope is critical during the condition of steady seepage.

Section of dam

Embankment dams are constructed of either earth fill or a combination of the earth and rock fill. Therefore, the embankment dams are basically built in areas where large amount of earth or rocks are available. They represent 75% of all dams in the world.

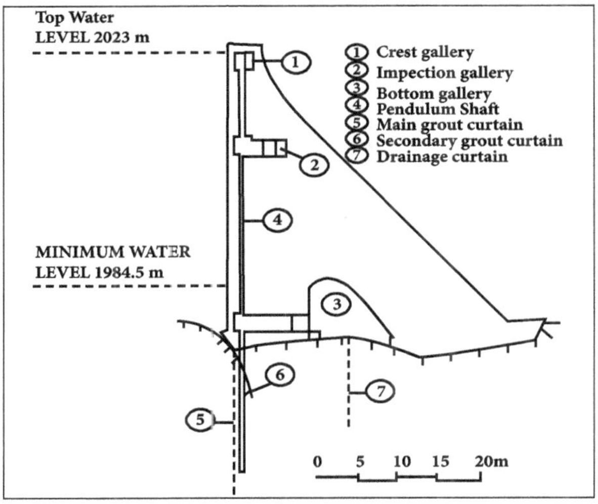

Gravity dam.

Gravity dams depends entirely on their own weight to resist the tremendous force of the stored water. In the earlier times, some dams were constructed with masonry blocks and concrete. Today, gravity dams are built by the mass concrete or by roller compacted concrete.

Arch dams are the concrete dams that curve the upstream towards the flow of water. They are generally built in narrow canyons, where the arch may transfer the water's force to canyon wall. Arch dams requires lesser concrete than gravity dams of the same length, but they require a solid rock foundation in order to support their weight.

Buttress dams depend for support on a series of vertical supports known as the buttresses, which run along the downstream face.

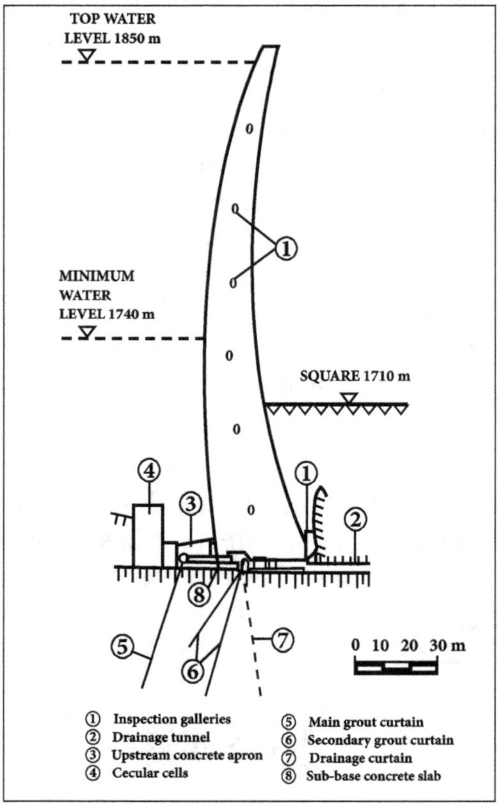

Arch dam.

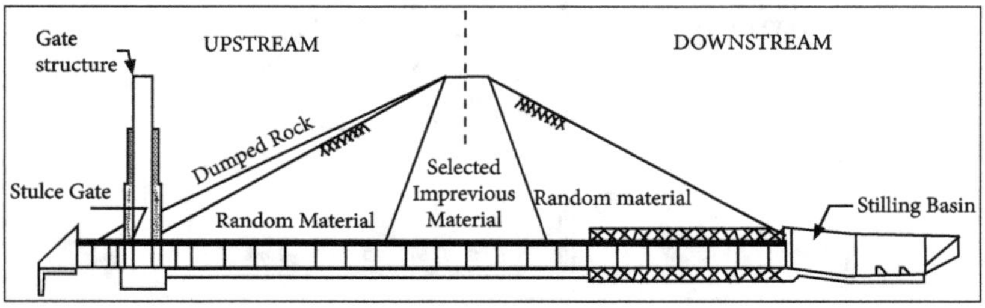

Embankment dam.

6.1.2 Criteria for Safe Design, Seepage and Measures for Control of Seepage-filters

- A fill of sufficiently low permeability should be developed out of the available materials, so as to best serve the intended purpose with minimum cost. Borrow pits should be as close to the dam site as possible, so as to reduce the leads.
- Sufficient spillway and outlets capacities should be provided so as to avoid the possibility of overtopping during design flood.
- Sufficient freeboard must be provided for wind set-up, wave action, frost action and earthquake motions.
- The seepage line (i.e. phreatic line) should remain well within the downstream face of the dam, so that no sloughing of the face occurs.
- There is little harm in seepage through a flood control dam. If the stability of foundations and embankments is not impaired, by piping, sloughing, etc.: But a conservation dam must be as watertight as possible.
- There should be no possibility of free flow of water from the upstream to the downstream face.
- The upstream face should be properly protected against wave action, and the downstream face against rains and-against waves up to tail water Provisions of horizontal berms at suitable intervals in the dis face may be thought of, so as to reduce the erosion due to flow of rain water. Ripraps should be provided on the entire u/s slope and als6 on the d/s slope near the toe and up to slightly above the tail water so as to avoid erosion.
- The portion of the downstream of the impervious core should be properly by providing horizontal filter drain or toe drain, or chimney drain, etc.
- The upstream and downstream should be so designed as to be stable under worst conditions of loading. These critical occur for u/s conditions during sudden drawdown of the reservoir and for the d/s slope during steady seepage under full reservoir.
- The u/s and d/s slope should be flat enough, as to provide sufficient base Width at the foundation level, such that the maximum shear stress developed remains well below the corresponding maximum shear strength of the soil, so as to provide a suitable factor of safety.
- We know that the consolidation of the soil does not take place instantaneously when the compaction is done by external leadings. It takes place slowly as the excess pore water goes out and the load is transferred to the soil grains. In

coarse gravels, the void openings are large enough so as to permit rapid escape of confined water and air, and full compaction may occur before the construction is over. But in fine grained impervious soils, the consolidation is slow. It, therefore, becomes necessary in 'Such cases, as to provide an additional height of the fill. After consolidation, the the embank-ment will be of the desired height. Hence, a suitable allowance in the height of embank-ment (between 2 to 3% of dam height, determined by laboratory tests) must be made in fine grained soils so as to account for the consolidation that may take place upto years after construction. Dewatering the foundations may sometimes be used to accelerate the process of consolidation.

- Since the stability of the embankment and foundation is very critical during construction or even after construction (i.e. during the period of consolidation), due to development of excessive pore pressures and consequent reduction in shear strength of soil, the embankment slopes must remain safe under this critical condition also.

All the above criteria must be satisfied and accounted for, in order to obtain a safe design and construction of an earth dam.

Controlling Seepage

Regularly scheduled monitoring and inspection is essential to detect seepage and prevent dam failure. Inspections must be made periodically throughout the year. Frequency must be based on the hazard classification of dam. Higher classified dams are need to be checked more common, compared to those that are classified under lower hazard. At the minimum all dams are to be visually inspected at least for every six months, before a predicted major storm event, during or after severe rainstorms or snowmelts, and are inspected weekly after the construction is complete and reservoir filling is ongoing, and for at least two months after the reservoir has been filled. Dam inspections performed on a regular basis are the most economical aid a dam owner may use to assure the safety and long life of the structure while reducing liability risks. If seepage is detected on a dam embankment or foundation, it must be closely monitored on a regular basis until it is corrected.

If seepage flow increases or embankment soils that shows signs of instability, corrective action must be taken quickly. The Seepage problems at high hazard dams has to be corrected immediately. If the problems are neglected and permitted to progress, they will result in loss of life and property downstream of the dam. A qualified geotechnical engineer or dam safety professional must be contacted for inspection and advice for all high hazard dam seepage problems. The type of controls deployed depends on source, type, and extent of seepage. Flow nets may also be installed as a method of studying the path that the seeping, or moving water follows. Then the correct action could be taken. If excessive water is flowing from soil piping or boils, or if the water is carrying sediment, an engineer or safety professional should be contacted for inspection and recommendations for further action. The reservoir level should be lowered when serious piping or embankment sliding/sloughing is occurring, and the cause of the condition corrected. The Sloughing and sliding due to the seepage at the toe of the embankment can be corrected by removing unstable soil and building a toe drain with filter out of permeable soil. Permeable soils and materials are commonly used at the toe of dams, to create drains to help them control and prevent the erosion. By using such materials, it allows the water to pass through without carrying the soil. The same thing may be done in locations where water is flowing but piping has not yet occurred.

Seepage, piping, and boils in the existing dams can be corrected, or slowed, by intercepting the water before it exits the downstream side of the dam. Some methods of intercepting include impermeable upstream blankets, cutoff trenches in the embankment, grout curtains, relief wells, and toe drains. Impermeable upstream blankets, or liners, are the most effective method, but they need complete drawdown of the reservoir. These blankets consist of low-permeability soil or a synthetic geo-membrane. The blankets may be deployed on the floor of the reservoir to prevent the foundation seepage. All cracks and erosion rills on the embankment must be filled, re-graded, and re-seeded. Borrowing rodents must be eliminated from dams, and any damage created need to be repaired by backfilling with a soil or filtered drain.

Safety Measures

The dam safety can be ensured if the following aspects are taken care of:

Hydraulic Failure

Such type of failure can be averted by giving:

- Adequate spillway capacity.
- Adequate freeboard in order dam safety is not endangered by overtopping during high floods.
- Proper maintenance of gates to make sure that they are always operative and do not get clogged.

Seepage Failure

Seepage failure can be corrected by:

- Providing filter at the toe to minimize the movement of the material.
- Seepage line is well within the body of the dam.

- Provision of the settlement, after the composition of the dam, is made from a normal 1 % of height to a maximum of 6 %.

Structural Failure

In spite of best geological and foundation investigations done prior to dam construction, geological problems may arise such as induced seepage, earth tremors, slides, gougy seams and sloughing in the vicinity of dam and reservoir area surface during the construction or several years after the reservoir filling.

Periodic geotechnical inspection is essential for the early detection and resolution of potential problems, besides provision of adequate rip-rap and its maintenance and drawdown within permissible limits.

Earthquake Failure

Necessary provisions should be made in the design of a dam to account for additional forces due to earthquakes.

6.2 Stability Analysis

Stability of Downstream Slope During Steady Seepage

Critical condition for d/s slope occurs when the reservoir is full and percolation is at its maximum rate. The direction of seepage forces tend to decrease stability. In other words, the pore water pressure (U) acting on the soil mass below the saturated line reduces the effective stress responsible for mobilising shearing resistance. The factor of safety in this case is given by,

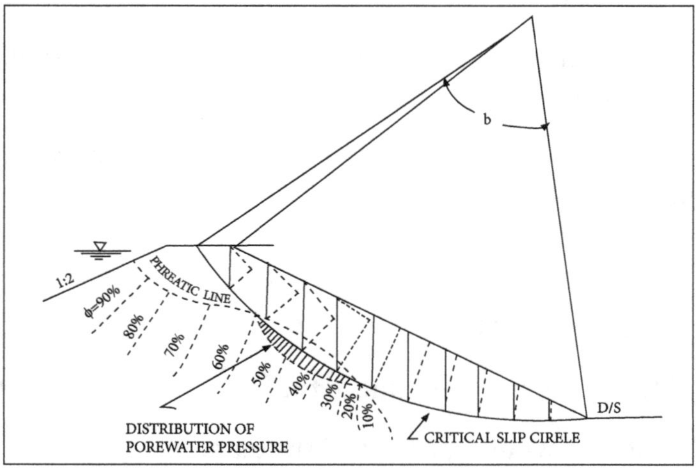

Stability Of d/s Slope During Steady Seepage.

$$F.S. = \frac{c\pounds + \tan\phi \sum(N-U)}{\sum T}$$

When $\sum U$ is the total pore pressure on the slip surface.

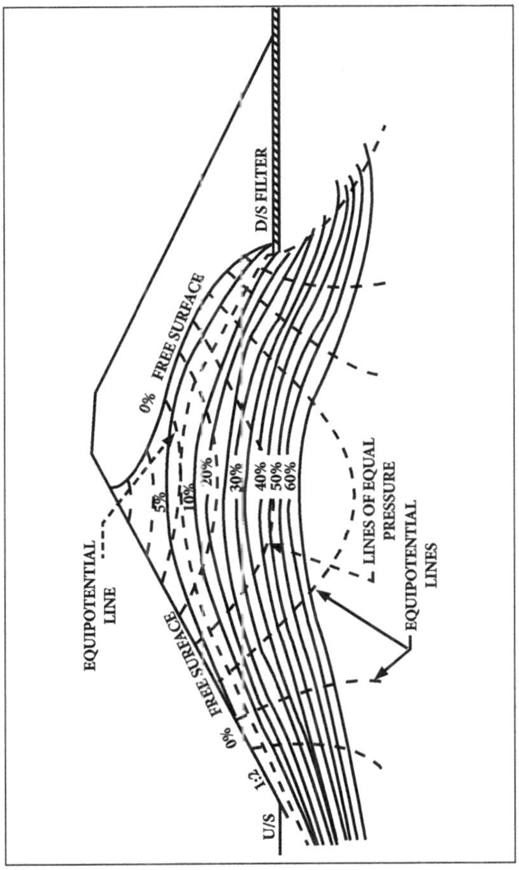

Solution With Tile Help Of Flow Net And Pressure Net.

The pore-water pressure at any point is represented by the piezometric head (h_w) at that point. Thus the variations of pore water pressure along a likely slip surface is obtained by measuring at each of its intersections with an equipotential line, the vertical height from that intersection to the level at which the equipotential line cuts the phreatic line. The pore pressure represented by vertical height so obtained are plotted to scale in a direction normal to the sliding surface at the respective point of intersection. The distribution of pore-water pressure on the critical slip surface during steady seepage is shown hatched in above figure. The area of U-diagram can be measured with the help of a planimeter, or else the rectangular plot method may be utilised.

If Au is the area of U-rectangle, then, $\sum U = A_U \times x^2$ where x is the scale of the drawing (i.e. 1 cm =x metres). In the absence of a flow net, the F.S. of the d/s slope can approximately be calculated from the equation,

$$F.S. = \frac{c£ + \tan\phi \Sigma N'}{\Sigma T}$$

Where the normal components N' are to be calculated on the buoyant unit weight (γ') of the dam, while T, components are to be calculated on the basis of saturated weight.

Stability of Upstream Slope During Sudden Drawdown

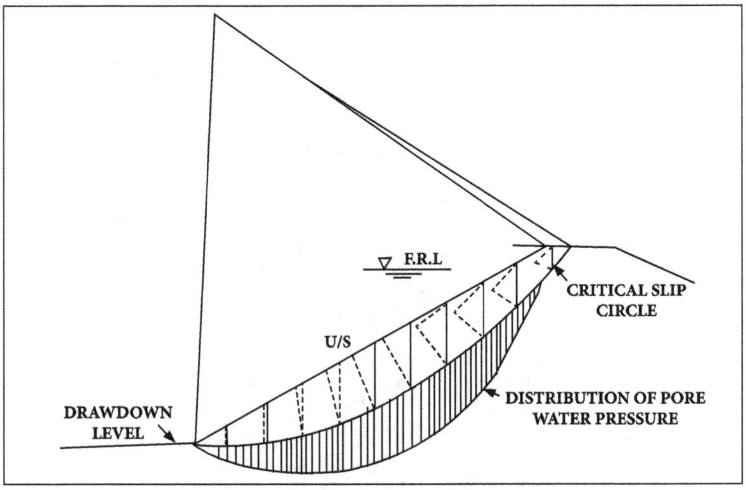

Stability Analysis During Drawdown.

For the u/s slope, steady seepage does not represent the critical state, because the seepage pressure then acts inwards from this slope and tends to increase the stability on the upstream side. For the u/s slope the critical condition is when the reservoir is suddenly emptied without allowing any appreciable change in the water level within the saturated mass of soil. This state is known as sudden drawdown. With complete drawdown, the hydrostatic force acting along the u/s slope at the time of full reservoir is removed. The weight of water within the soil now tends to help a sliding failure, without the hydrostatic pressure on the slope to counteract it. Or, by another interpretation, while the Disturbing force is the tangential component of the saturated weight the shear resistance is considerably reduced on account of development of pore water pressure on a likely slip surface. The effect of drawdown on slope stability varies appreciably with opportunity for drainage at the base. If the base is quite impervious, flow lines tend to be horizontal and in outward direction towards the slope, which is quite unfavourable to stability. On the other hand, if the base material is pervious, flow pattern tends to be downwards and this is more favourable with respect to stability.

The magnitude and distribution of pore water pressure on a likely slip surface is estimated from pressure net which is developed from the flow net. The above figure represents the pressure- net for sudden drawdown case. The pressure net gives line of equal piezometric heads (h_w) expressed in terms of percentage of total hydraulic heads H when the reservoir is full. At any point, we have,

$\varphi = h. + y$

Where φ = hydraulic potential at the point, expressed as percentage of total head.

h_w = piezometric head, expressed as percentage of total head H.

y =elevation head, expressed as percentage of total head H.

If we want to plot the 'line of equal pressure' corresponding to h_w =20 % (say), we have,

$$h_w = 20 = \Phi - y$$

If we want to plot the 'line of equal pressure' corresponding to h_w = 20 % (say), we have,

To find a point corresponding to h_w = 20 % on φ =30 %, we get from above equation, y=30-20 =10 % of H. Thus the point is located. Similarly, to get another point of h_w = 20 % on φ =40%, y=40 - 20=20 % of H and so on. The various points so obtained may be joined by a smooth curve to get a contour for h_w = 20 %. In this way, a pressure net may be prepared. In order to plot the distribution of pore water pressure on the assumed slip surface, the pressure net of above figure is superimposed on the diagram showing the location of slip surface. At the intersection points of the slip surface with the contours of pressure net, radial ordinates of the values of h_w's of corresponding contours are erected. The curve joining the ends of these ordinates give the distribution of pore water pressure, as illustrated in Figure below. The F.S. is then calculated by the expression,

$$\text{F.S.} = \frac{c\pounds + \tan\phi \sum (N - U)}{\sum T}$$

Approximate Analysis

In the absence of the flow net the factor of safety can be calculated from the expression,

$$\text{F.S.} = \frac{c\pounds + \tan\phi \sum N'}{\sum T}$$

In which the N' components are computed with respect to the submerged density (γ') of the u/s slope, while T components are calculated with respect to the saturated density.

Analysis for Sloughing Of Upstream Slope During Sudden Drawdowon

This is another approximate method for finding the factor of safety against sliding or sloughing u/s slope during sudden drawdown.

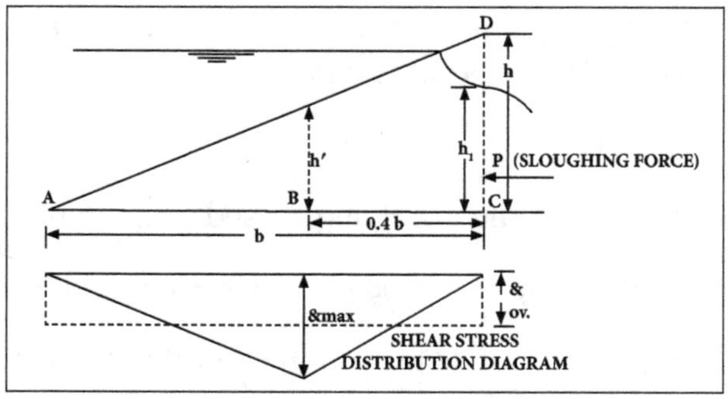

Sloughing of Ws Slope During Sudden Drawdown.

Let,

γ' =submerged unit weight of the material in u/s portion of dam.

γ_s =saturated unit weight of the material in u/s portion of dam.

b =horizontal length of the slope AD.

p =Total horizontal shear or sloughing force on the upstream portion of dam.

s_a =Average unit shear over upstream part of dam.

s_{max} =Maximum unit shear over upstream part of dam.

h, h_1, h' =heights as marked in Figure.

Then the horizontal shear force is given by,

$$p = \frac{y_s h^2}{2} \tan^2\left(45^\circ - \frac{\phi}{2}\right) + \frac{y_s h_1^2}{2}$$

$$S_{\alpha v} = \frac{p}{b}$$

And,

$$S_{max} = 2S_{\alpha v} = \frac{2P}{b}$$

The maximum shear occurs at a distance of 0.4 b from the u/s point, as shown.

Where,

$$N_e = y'A$$

And,

$$A = \text{area of u/s slop} = \frac{1}{2} bh$$

$$\therefore \text{ Factor of safety} = \frac{R}{P}$$

$$R = \text{shearing resistance} = N_e \tan\phi + cb$$

His should be greater than 2.

(Note. If the analysis is done by slip circle method, factor of safety may be taken as 1.5.)

The above factor of safety is with respect to average shear. In order to find the factor of safety with respect to maximum shear, we have,

$$S_{max} = 2S_{av} = \frac{2P}{b} (\text{at point B})$$

$$\begin{aligned} S_r &= \text{Shearing resistance at B} \\ &= h'y' \tan\phi + c \\ &= 0.6 h'y' \tan\phi + c \end{aligned}$$

Therefore F.S at the point of maximum shear,

$$= \frac{S_r}{S_{max}}$$

This should be greater than 1.

6.3 Spillways: Types

The spillways are openings provided at the body of the dam to discharge the excess water or flood water safely when the water level rises above normal pool level.

Necessity of Spillways

- The height of dam is fixed according to the maximum reservoir capacity. The normal pool level is the one that indicates the maximum capacity of the reservoir. The water is never stored in reservoir above this level. The dam might fail by overturning so for the safety of the dam the spillways are essential.
- The top of the dam is usually utilized by making road. The surplus water is not to be allowed till the dam reaches its top, so to stop the over topping by the surplus water, the spillways become extremely responsible.
- To protect the downstream base and floor of the dam from the effect of scouring and erosion, the spillways are provided in order to make the excess water flow smoothly.

Spillway Components

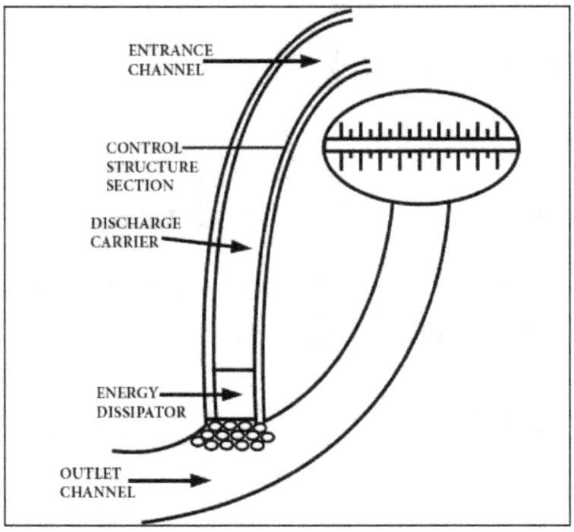

In general spillways comprise five distinct components namely:

- A discharge carrier.
- An entrance channel.
- A control structure.
- An outlet channel.
- An energy dissipator.

The entrance channel transfers water from reservoir to the control structure which regulates discharge from the reservoir. Water is then conveyed from reservoir to low-level energy dissipator on riverbed by the discharge conveyor. An energy dissipator is needed to reduce the high velocity of the flow to a non-scouring magnitude.

Factors Affecting type and Design of Spillways

Safety considerations consistent with economy: Many failures of dams have resulted from the improperly designed spillway or spillways of inadequate capacity. The Properly designed structure of the adequate capacity can be found to be only moderately higher in cost than a structure of inadequate capacity.

Hydrological and site conditions: The spillway design and its capacity depend on.

Inflow discharge, its frequency, and shape of the hydrograph:

- Height of the dam.
- Capacity curve.
- Geological and other site conditions Important topographical features, which affect spillways design.

- Steepness of the terrain.
- Amount of the excavation and possibility of its use as the embankment material.
- The possibility of the scour.
- Stability of slopes, safe bearing capacity of the soils.
- Permeability of the soils.

For example, in case of narrow valley dams, side or chute channel spillway is very seldom possible, because of steepness of the banks and their insufficient stability.

Types of Spillway

Spillway is usually referred to as controlled or uncontrolled depending on whether spillway gates for controlling the flow have been provided or not. A free or uncontrolled spillway automatically releases water whenever the reservoir level rises above the overflow crest level.

The main advantage of an uncontrolled spillway is that it does not require constant attendance and operation of the regulating devices by an operator. Besides, there are no problems related to the maintenance and repair of the devices. If it is not possible to provide a sufficiently long uncontrolled spillway crest or obtain a large enough surcharge head to meet the requirements of spillway capacity, one has to provide regulating gates. Such gates enable release of water, if required, even when the reservoir level is below the normal reservoir water surface level. Most common types of spillway are as follows:

- Free overfall (straight drop) spillway.
- Ogee (overflow) spillway.
- Side-channel spillway.
- Chute (or open channel or trough) spillway.
- Shaft (or morning glory) spillway.
- Siphon spillway.
- Cascade spillway.
- Tunnel (conduit) spillway.

Free Overfall Spillway

As the name indicates, the flowing water drops freely from the crest of a free over fall spillway. At times, the crest is extended in the form of an overhanging lip to direct small discharges away from the downstream face of the overflow section. The underside of the falling water jet is properly ventilated so that the jet does not pulsate. Such a spillway is better suited for a thin arch dam whose downstream face is nearly vertical. Since the flowing water usually drops into the stream bed, objectionable scour may occur in some cases and a deep plunge pool may be formed. If erosion cannot be tolerated, plunge pool is created by constructing an auxiliary dam downstream of the main dam. Alternatively, a basin is excavated and is provided with a concrete apron. When tail-water depth is sufficient, a hydraulic jump forms when the water jet falls upon a flat apron. Free over fall spillways are restricted only to situations where the hydraulic drop from the reservoir level to tail-water level is less than about 6 m.

Ogee (Overflow) Spillway

An ogee spillway has a control weir whose profile is as shown in figure. The upper part of the spillway surface matches closely with the profile of the lower nappe of a ventilated sheet of water falling freely from a sharp-crested weir. The lower part of the spillway surface is tangential to the upper curve and supports the falling sheet of water. The downstream end of the spillway is in the form of a reverse curve which turns the flow into the apron of a stilling basin or into the spillway discharge channel. An ogee spillway is generally used for concrete and masonry dams. It is ideally suited to wider valleys where sufficient crest length may be provided.

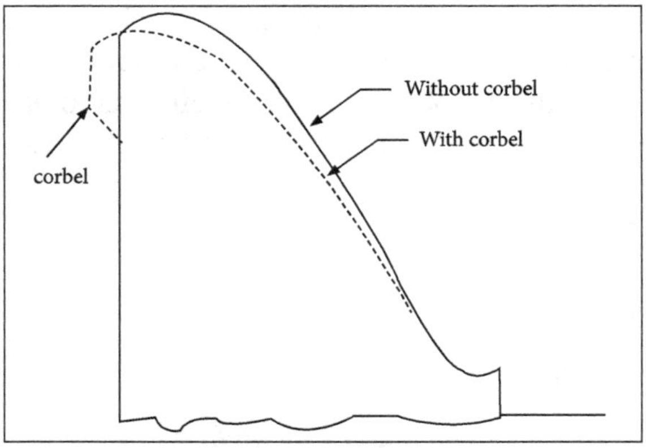

Ogee (Overflow) Spillway.

The profile of an ogee spillway is designed for a given design discharge (or corresponding design surcharge head or, simply, design head). When the flowing discharge equals the design discharge, the flow adheres to the spillway surface with minimum interference from the boundary surface and air has no access to the underside of the water sheet. The discharge efficiency is maximum under such condition and the pressure along the spillway surface is atmospheric. If the flowing discharge exceeds the design discharge, the water sheet tends to pull away from the spillway surface and thus produces sub-atmospheric pressure along the surface of the spillway. While negative pressure may cause cavitation and other problems, it increases the effective head and increases the discharge. On the other hand, positive hydrostatic pressure will occur on the spillway surface, if the flowing discharge is less than the design discharge.

Model tests have indicated that the design head may be safely exceeded by about 50%beyond which cavitation may develop. Therefore, spillway profile may be designed for 75% of the peak head for the maximum design flood.

An upstream overhang, known as corbel, is added to the upstream face of the spillway as shown in Fig. 17.6. The effect of the corbel is to shift the nappe (and, hence, the spillway profile) backward which results in saving of concrete. If the height of the vertical face of the corbel is kept more than 0.3 times the head over the crest, the discharge coefficient of the spillway will be practically the same as it would be if the vertical face of the corbel were to extend to the full height of the spillway.

Side-channel Spillway

The control weir of a side-channel spillway is located alongside and approximately parallel to the upstream portion of the spillway discharge channel which itself may be either an open channel, a closed conduit, or an inclined tunnel. The spillway discharge flows over the weir crest and falls into a narrow trough (i.e., upstream of the discharge channel) and takes an approximately 90°-turn before continuing into the spillway discharge channel. The control structure, in plan, may be straight, curved, semi-circular or U-shaped. The overflow section may be broad-crested instead of ogee-shaped.

One should note that when the flow in the side channel trough is subcritical, the incoming low from the control structure (i.e., the overflow crest) will not cause high transverse velocities because of the low drop due to relatively higher depth of flow in the trough. This would effect good diffusion and intermingling of the incoming flow with the trough water due to relatively low velocities of both the incoming flow and the flow in the trough. Therefore, there would be comparatively smooth flow in the side channel trough. However, when the flow in the side channel trough is supercritical, the flow velocities in the trough would be high and the depth of flow small, causing the incoming flow to have a relatively higher drop. Therefore, the intermixing of the high-energy transverse flow with the trough stream will be rough and turbulent producing violent wave action causing vibrations. Therefore, the flow in the side channel trough should be maintained at subcritical condition for good hydraulic performance.

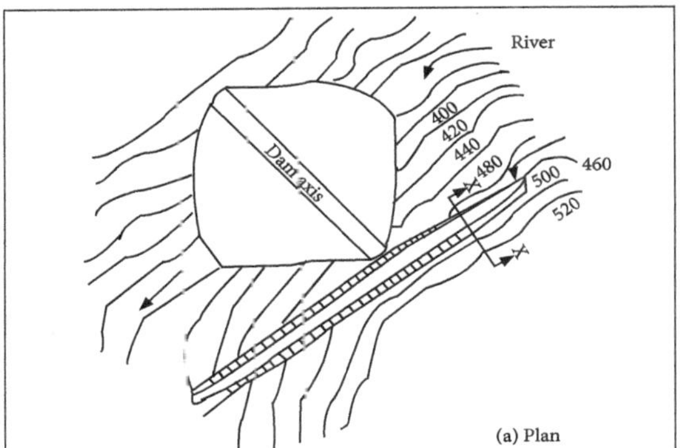

(a) Plan

Moreover, the amount of excavation would also increase for larger bed widths. While the flow in the side channel trough should preferably be subcritical, the flow in the discharge channel is supercritical and a control section downstream of the trough is provided by either constricting the channel width or raising the channel bottom.

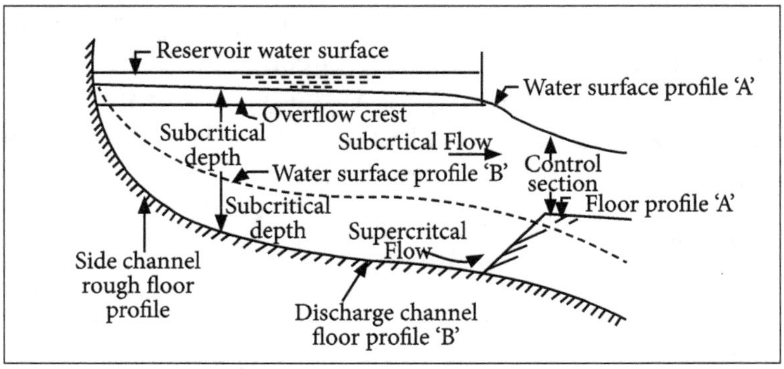

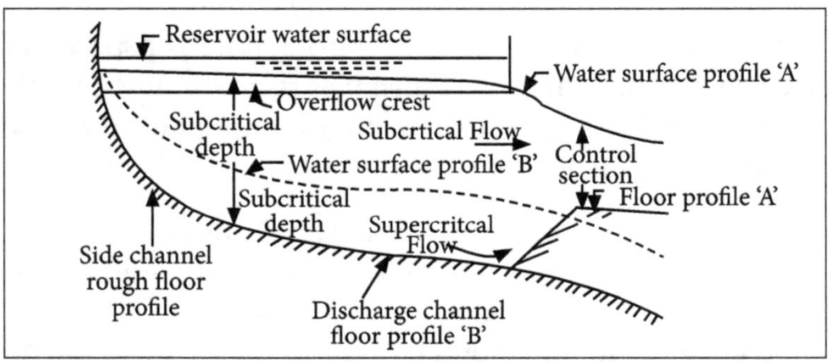

Side-Channel Spillway.

Chute Spillway

In a chute (or trough) spillway, the spillway discharge flows in an open channel (named as 'chute' or 'trough') right from the reservoir to the downstream river. The open channel can be located either along the abutment of the dam or through a saddle. The channel bed should always be kept in excavation and its side slope must be designed to be stable with sufficient margin of safety. As far as possible, bends in the channel should be avoided. If it becomes necessary to provide a bend, it should be gentle. The spillway control structure can be an overflow crest, or a gated orifice or some other suitable control device. The control device is usually placed normal or nearly normal to the axis of the chute. The simplest form of a chute spillway is an open channel with a straight centre line and constant width. However, often the axis of either the entrance channel or the discharge channel is curved to suit the topography of the site. The flow condition varies from subcritical upstream of the controlling crest to critical at the crest and supercritical in the discharge channel. The chute spillway is ideally suited with earth fill dams because of: (i) simplicity of their design and construction, (ii) their adaptability to all types of foundation ranging from solid rock to soft clay, and (iii) overall economy usually obtained by the use of large amounts of spillway excavation for the construction of embankment. The chute spillway is also suitable for concrete dams constructed in narrow valleys across a river whose bed is erodible for which the ogee spillway becomes unsuitable.

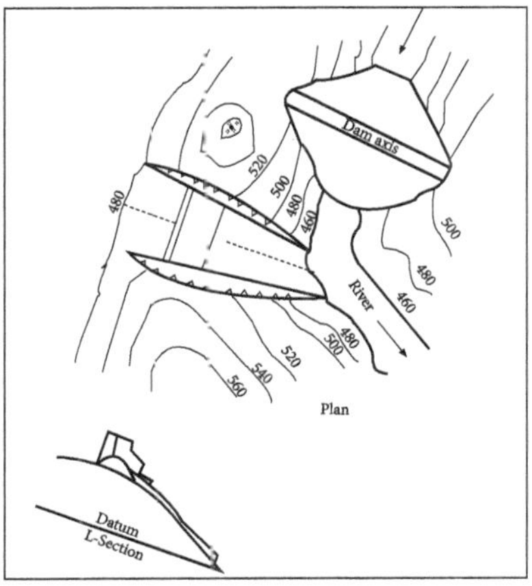

Chute Spillway.

Shaft Spillway

In a shaft spillway water enters a horizontal crest, drops through a vertical or sloping shaft, and then flows to the downstream river channel through a horizontal or nearly horizontal conduit or tunnel. A rock outcrop projecting into the reservoir slightly upstream of the dam would be an ideal site for shaft spillway. Depending upon the level of the rock out crop and the required crest level, a spillway may have to be either constructed or excavated. The diversion tunnels, if used for river diversion purposes during construction, can be utilized for discharge tunnels of the spillway. Radial piers provided on the spillway crest ensure radial flow towards the spillway and also provide support to the bridge which would connect the spillway with the dam or the surrounding hill.

A shaft spillway with a funnel-shaped inlet is called a "morning glory" or "glory hole" spillway. One of its distinguishing characteristics is that near maximum capacity of the spillway is attained at relatively low heads. Therefore, a shaft spillway is ideal when maximum spillway discharge is not likely to be exceeded. Because of this feature, however, the spillway becomes unsuitable when a flow larger than the selected design flow occurs. This disadvantage can begot rid of by providing an auxiliary or emergency spillway and using the shaft spillway as service spillway.

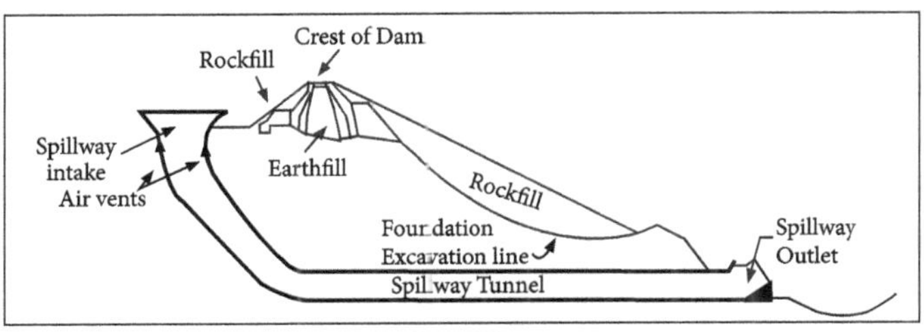

Shaft Spillway.

Siphon Spillway

A siphon spillway is essentially a closed conduit system which uses the principle of siphonic action. The conduit system is the shape of an inverted U of unequal legs with its inlet end at or below normal reservoir storage level. When the reservoir water level rises above the normal level, the initial flow of water is similar to the flow over a weir. When the air in the bend has been exhausted, siphonic action starts and continuous flow is maintained until air enters the bend. The inlet end of the conduit is placed well below the normal reservoir water level to prevent ice and drift from entering the conduit. Therefore, once the siphonic action starts, the spillway continues to discharge even after the reservoir water level falls below the normal level. As such, a siphon-breaking air vent is always provided so that siphonic action can be broken once the reservoir water level has been drawn down to the normal level in the reservoir. Siphon spillways can be either constructed of concrete or formed of steel pipe. The thickness of the wall of the siphon structure should, however, be sufficiently strong to withstand the negative pressures which develop in the siphon. Pressure at the throat section can be determined by the use of Bernoulli's equation.

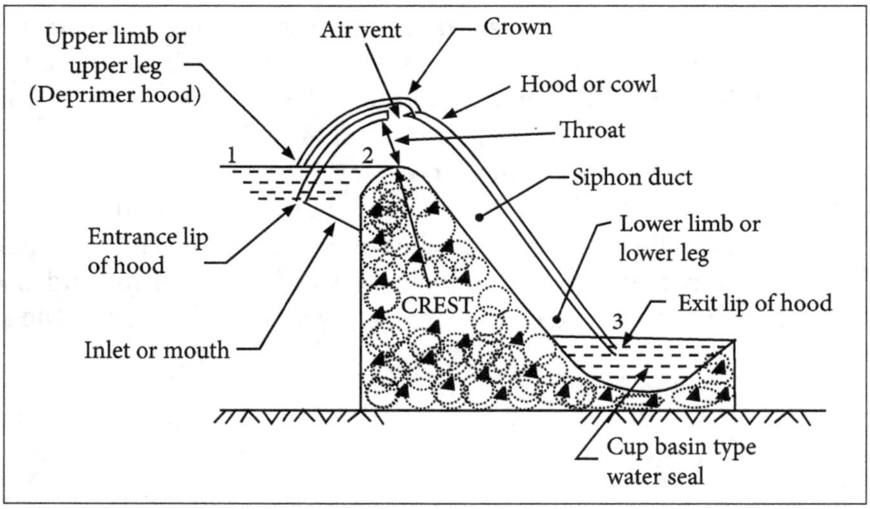

Siphon Spillway.

Cascade Spillway

In case of very high dams the kinetic energy at the toe of the dam will be very high and the tailwater depth in the river may not be adequate for a single-fall hydraulic jump or roller bucket stilling basin. Narrow and curved canyons consisting of fractured rock would not be suitable for trajectory buckets. In such situations, especially for high earth and rockfill dams for which spillway is a major structure, possibility of providing a cascade of falls with a stilling basin a teach fall as shown in figure below must be considered. The cascade spillway (2, 8) is likely to be an ideal choice for a high rock fill dam for which the material has been obtained from a quarry located

downstream of the dam so that the flood waters may be discharged over the quarry face. As the quarry would usually be excavated in benches, they may as well form the steps in the cascade. A cascade spillway has been planned at the proposed 218 m high Tehri dam on the Bhagirathi River in the Ganga valley of the central Himalayas. At Darmouth dam in Australia, the spillway to pass 2700 m^3/s of flood discharge is an unlined cascade in granite. The benches of the cascade will be 5 m high and of varying widths to suit the topography of the site. Although, a cascade spillway would be an attraction during floods, it may not be always acceptable for environmental reasons which demand that the quarries be always located upstream of the dam and below the normal water surface level of the reservoir so as to cause minimum disfigurement of the land.

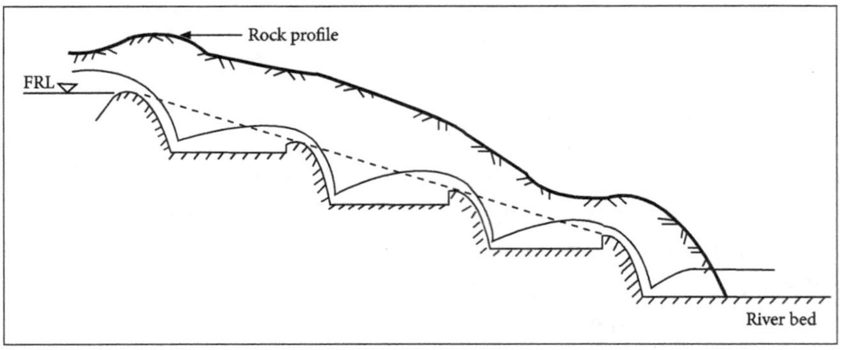

Cascade spillway.

Tunnel Spillway

A tunnel spillway discharges water through closed channels or tunnels laid around or under a dam. The closed channels can be in the form of a vertical or inclined shaft, a conduit constructed in an open cut and back-filled with earth materials, or a horizontal tunnel through earth or rock. In narrow canyons with steep abutments as well as in wide valleys with abutments faraway from the stream channel, tunnel spillways may prove to be advantageous. In such situations, the conduit of the spillway can be easily located under the dam near the stream bed.

6.3.1 Design Principles of Ogee Spillways

The ogee spillway is the modified form of drop spillway. The downstream profile of the spillway is made to coincide with shape of the lower nappe of the free falling waterjet from the sharp crested weir. In this case, the shape of the lower nappe is similar to a projectile and hence downstream surface of the ogee spillway will have to follow the parabolic path where “o” is the origin of parabola. The downstream face of the spillway forms concave curve from a point “T” and meets with the downstream floor. This point “T” is called as point of tangency. Therefore, the spillway takes the shape of the letter “S” .Thus, this spillway is termed as ogee spillway.

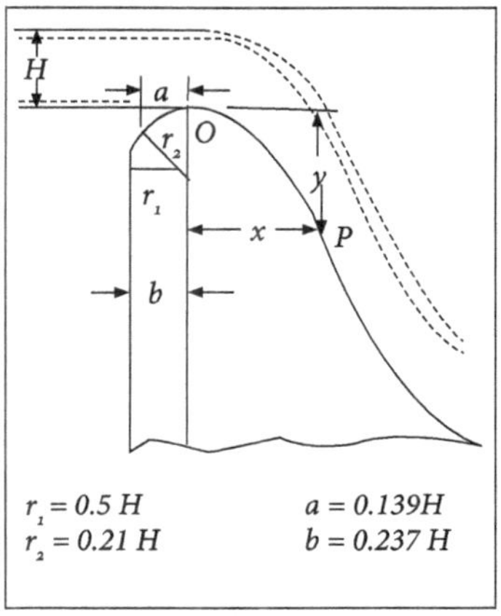

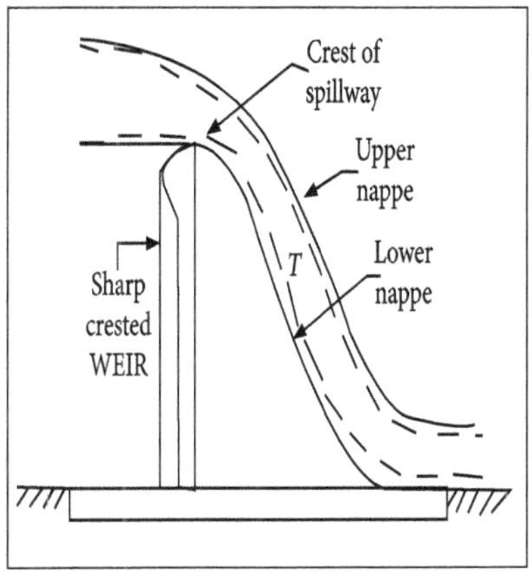

Ogee spillway.

Shape of the Ogee Spillway

The shape of the lower nappe is not the same for all the head of water above the crest of the weir. It differs with head of water. But for design of the ogee spillway the maximum head is considered. If the spillway runs with the maximum head, then the overflowing water just follows the curved profile of the spillway and there will be no gap between the water and the spillway surface and the discharge is maximum.

When the actual head becomes more than the designed head, the lower nappe does not follow ogee profile and gets separated from spillway surface. Hence, a negative pressure develops at the point of the separation. Due to the negative pressure, air bubbles are formed within the flowing water. These air bubbles air responsible for frictional force which causes much damage to spillway surface.

Again, if the head of water is less than designed head, the waterjet adheres to the body of the spillway and creases positive pressure which reduces the discharge through the spillway.

The shape of the ogee spillway has been developed by U.S Army Corps Engineers which is called as "Water-way experimental station spillway shape". The equation given by them is, $X^n = K \times H^{n-1} \times Y$, where, x and y are the coordinates of a point P on the ogee profile taking O as origin. K and n are the constants according to the slope of the upstream face of spillway.

Shape of u/s face of spillway	**K**	**N**
Vertical	2.0	1.85
1:3 (H:V)	1.936	1.836
$1:1^{1/2}$ (H:V)	1.939	1.810
1:1 (H:V)	1.873	1.776

Therefore, for different values of y, the values of x are determined considering the slope of u/s face. The value of r_1, r_2, a and b are determined. The results are tabulated for the constructional guidance.

Problem

Let us design and sketch the shape of an Ogee type spillway for the following data using the empirical equation developed by US Army Corps Engineers.

Upstream Head, H = 20 m

Shape of the upstream face = 1: 1½ (H:V)

Values of K and n are 1.939 and 1.81 respectively.

Solution:

X (m)	Y (m)
5	0.84
10	2.94
15	6.13
20	10.31
25	15.44

$r_1 = 0.5H = 0.5\times20 = 10$ m

$r_2 = 0.21H = 0.21\times20 = 4.2$ 4.5 m

$a = 0.139H = 0.139\times20 = 2.78$ 3 m

$b = 0.237H = 0.237\times20 = 4.74$ 5m

We know,

$$X^n = K \times H^{n-1} \times Y$$

$$X^{1.81} = Y\times1.939 \times (20)^{1.81-1}$$

$$X^{1.81} = Y\times21.95$$

$$Y = 0.159 \text{ m}$$

From the different values of X, the values of Y are found which is given below in a tabulated form: Finally the shape of Ogee type of spillway,

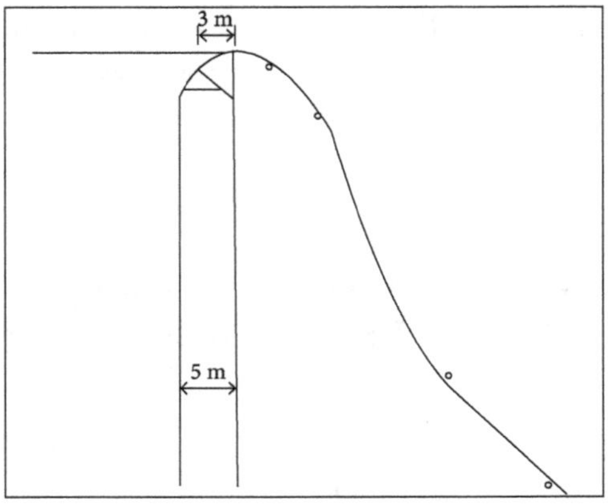

Free or uncontrolled overflow crest of a spillway is the simplest form of control as it automatically releases water whenever the reservoir water level rises above the crest level. For such crests, there is no need of constant attendance and regulation of the control devices by an operator.

Besides, the problems of maintenance and repair of the controlling device also do not arise. However, when sufficiently long uncontrolled crest or large surcharge head for the required. Spillway capacity cannot be obtained, a regulation gate may be necessary. Such regulating devices enable the spillway to release storages even when the water level in the reservoir is below the normal reservoir water surface. Gates can be provided on all types of spillways except the siphon spillway. The installation of gates involves additional expenditure on initial cost, and on their repair and maintenance. The selection of type and size of the controlling device depends on several factors, such as:

- Discharge characteristics of the device.
- Climate.
- Frequency and nature of floods.
- Winter storage requirements.
- The need for handling ice and debris.
- Special operating requirements such as presence of operator during periods of flood, the availability of electricity, operating mechanism, and so on. In addition, economy, reliability, efficiency, and adaptability of the regulating device must also the looked into.

Types of Spillways Crest Gates

The following types of regulating devices are generally used:

- Flashboards and stoplogs.
- Rectangular lift gates.

- Radial gates.
- Drum gates.

These may be controlled either manually or automatically through mechanical or hydraulic operations.

Flashboards and Stoplogs

Flashboards and stoplogs raise the reservoir storage level above a fixed spillway crest level when the spillway is not required to release flood. The flashboards usually consist of individual boards or panels of 1.0 to 1.25 m height. These are hinged at the bottom and are supported against water pressure by struts. Stoplogs are individual beams or girders set one upon the other to form a bulkhead supported in grooves at each end of the span. To increase the spillway capacity, the flashboards or stoplogs are removed prior to the flood. Alternatively, they are designed and arranged so that they can be removed while being overtopped. Flashboards and stop logs are simple and economical type of regulating devices which provide an unobstructed crest when removed. However, they have the following disadvantages:

- They present a hazard if not removed in time to pass floods, especially where the reservoir area is small and the stream is subject to flash floods.
- They require attendance of an operator or crew to remove them, unless designed to fall automatically.
- Ordinarily they cannot be restored to position while water flows over the crest.
- If they are designed to fail when the water reaches certain stage, their operation is uncertain, and when they fail they release sudden and undesirably large outflows.
- If the spillway functions frequently, the repeated replacement of flash boards may be costly.

Vertical Lift Gates

These are usually rectangular in shape and made of steel which span horizontally between guide grooves in supporting piers and move vertically in their own plane. The gates are raised or lowered by an overhead hoist and water is released by undershot orifice flow for all gate openings. Sliding gates offer large sliding friction due to water pressure and, therefore, require a large hoisting capacity. The use of wheels (along each side of the gate) would reduce the amount of sliding friction and thereby permit the use of a smaller hoist. Vertical lift gates have been used for spans and heights of the order of 20 m and 15 m, respectively. At larger heights, however, the problem of a raised operating platform becomes important.

Radial Gates

These are made of steel plates which form a segment of a cylinder which itself is attached to supporting bearing by radial arms. The cylindrical plate is kept concentric to the supporting pins so that the entire thrust of the water load passes through the pins and only a small amount of moment needs to be overcome in raising or lowering the gate. The hoisting loads then include only the weight of the gate, the sliding friction, and the frictional resistance at the pins. The small hoisting effort required for the operation of the radial gates makes hand operations at small installations possible. Besides, they require lesser head rooms than required by vertical lift gates. All these advantages make the radial gates more adaptable.

Drum Gates

Drum gates are hollow (and, therefore, buoyant), triangular in section, and made of steel plates. The drum gate is hinged at the upstream lip of a hydraulic chamber in the weir structure in which the gate floats. Water introduced into or drawn from the hydraulic chamber causes the gate to swing upwards or downwards. The inflow or outflow of water to the chamber is governed by controls located in the piers adjacent to the chambers.

6.3.2 Energy Dissipation Below Spillways-stilling Basin and its Appurtenances

Energy Dissipators

Different types of energy dissipators may be used along with a spillway, alone or in combination of more than one depending upon the energy to be dissipated and erosion control required downstream of a dam. Broadly the energy dissipators are classified under two categories – Stilling basins or Bucket Type. Each of these are further subcategorized as given below.

Stilling Basin Type Energy Dissipators

They may fundamentally be divided into two types,

1. Hydraulic Jump Type Stilling Basins

Horizontal apron type:

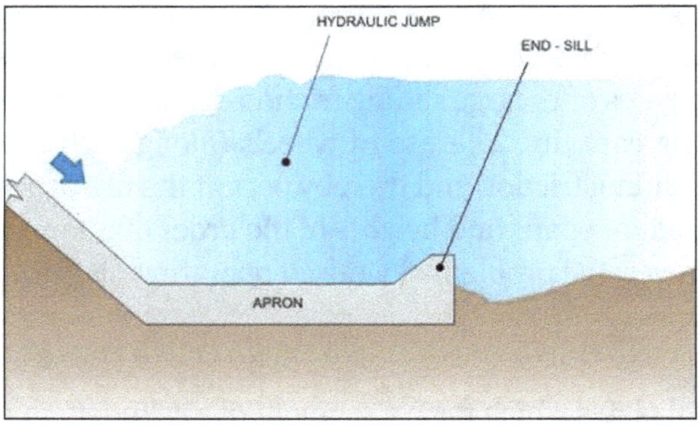

Sloping apron type:

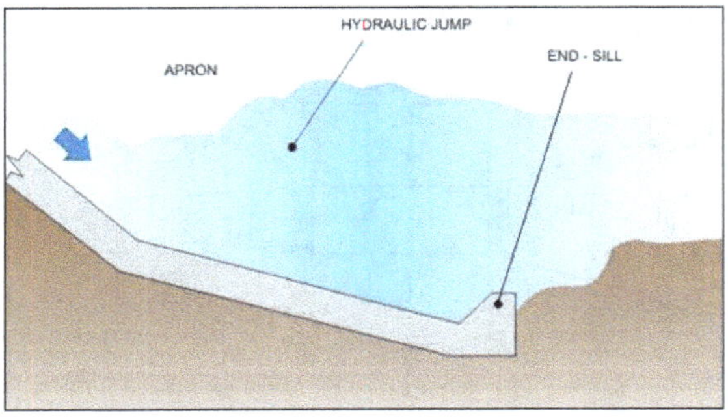

2. Jet Diffusion Type Stilling Basins

Jet diffusion stilling basins:

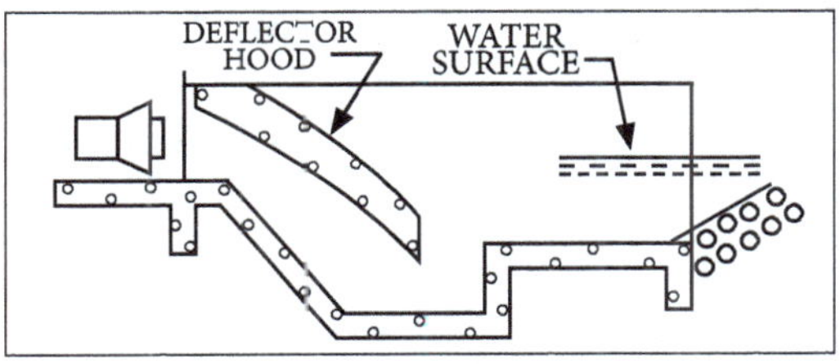

Interacting jet dissipators:

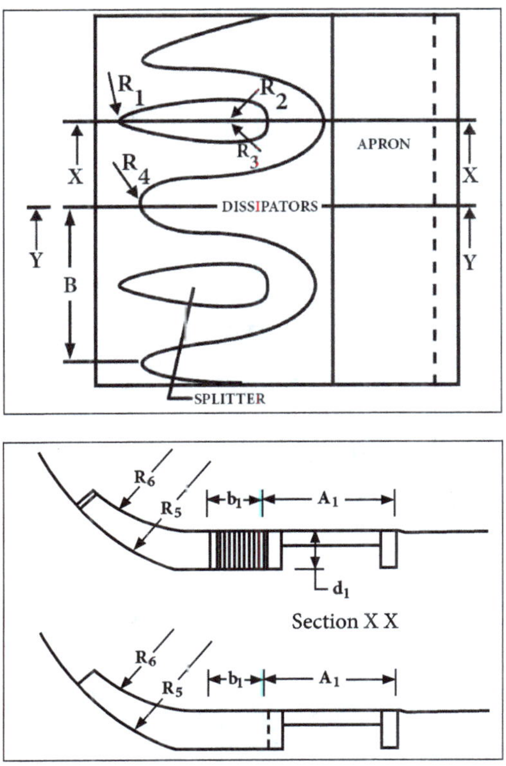

Free jet stilling basins:

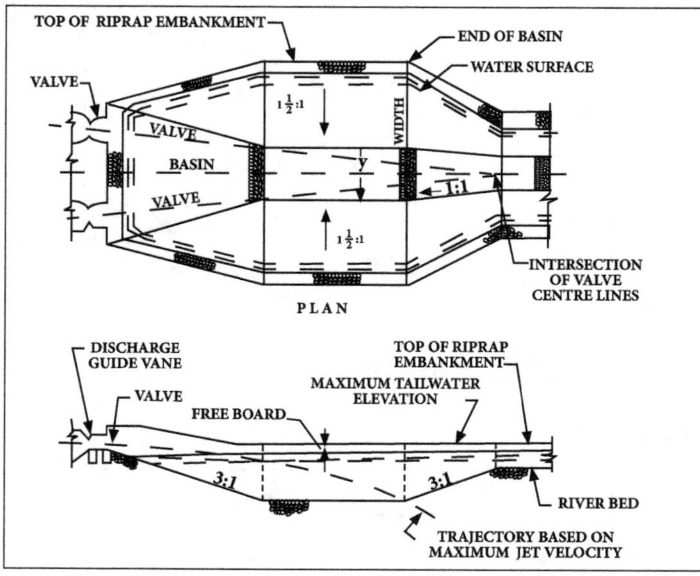

Hump stilling basins:

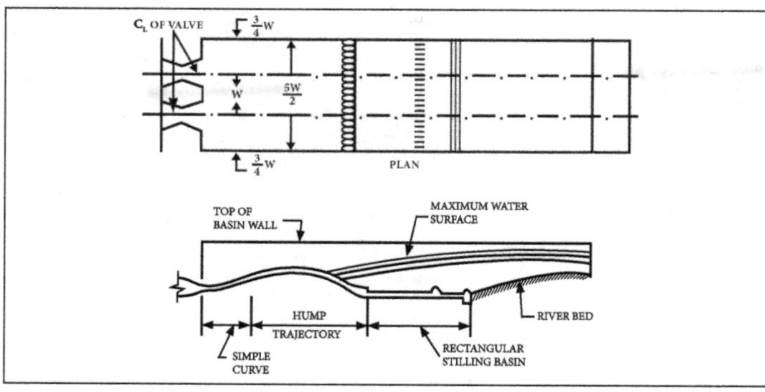

Impact stilling basins:

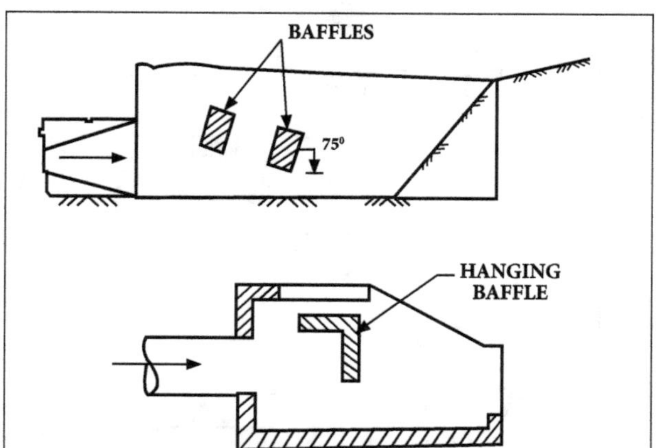

Bucket Type Energy Dissipators

This type of energy dissipators includes the following:

- Solid roller bucket.
- Slotted roller bucket.
- Ski jump (Flip/Trajectory) bucket.

The shapes of the different types of bucket-type stilling basins have been given in section. Usually the hydraulic jump type stilling basins and the three types of bucket-type energy dissipators are commonly used in conjunction with spillways of major projects. The detailed designs of these are dealt in subsequent sections.

Since energy dissipators are an integral part of a dam's spillway section, they have to be viewed in conjunction with the latter. Two typical examples have been shown in Figures though it must be remembered that any type of energy dissipator may go with any type of spillway, depending on the specific site conditions.

Permissions

All chapters in this book are published with permission under the Creative Commons Attribution Share Alike License or equivalent. Every chapter published in this book has been scrutinized by our experts. Their significance has been extensively debated. The topics covered herein carry significant information for a comprehensive understanding. They may even be implemented as practical applications or may be referred to as a beginning point for further studies.

We would like to thank the editorial team for lending their expertise to make the book truly unique. They have played a crucial role in the development of this book. Without their invaluable contributions this book wouldn't have been possible. They have made vital efforts to compile up to date information on the varied aspects of this subject to make this book a valuable addition to the collection of many professionals and students.

This book was conceptualized with the vision of imparting up-to-date and integrated information in this field. To ensure the same, a matchless editorial board was set up. Every individual on the board went through rigorous rounds of assessment to prove their worth. After which they invested a large part of their time researching and compiling the most relevant data for our readers.

The editorial board has been involved in producing this book since its inception. They have spent rigorous hours researching and exploring the diverse topics which have resulted in the successful publishing of this book. They have passed on their knowledge of decades through this book. To expedite this challenging task, the publisher supported the team at every step. A small team of assistant editors was also appointed to further simplify the editing procedure and attain best results for the readers.

Apart from the editorial board, the designing team has also invested a significant amount of their time in understanding the subject and creating the most relevant covers. They scrutinized every image to scout for the most suitable representation of the subject and create an appropriate cover for the book.

The publishing team has been an ardent support to the editorial, designing and production team. Their endless efforts to recruit the best for this project, has resulted in the accomplishment of this book. They are a veteran in the field of academics and their pool of knowledge is as vast as their experience in printing. Their expertise and guidance has proved useful at every step. Their uncompromising quality standards have made this book an exceptional effort. Their encouragement from time to time has been an inspiration for everyone.

The publisher and the editorial board hope that this book will prove to be a valuable piece of knowledge for students, practitioners and scholars across the globe.

Index